SNOW AVALANCHES

Snow Avalanches

Beliefs, Facts, and Science

François Louchet

Professor Emeritus of Condensed Matter Physics at Grenoble University (Grenoble institute of Technology), now retired, 38410 St Martin d'Uriage, France; https://sites.google.com/site/flouchet/, francoislouchet38@gmail.com

OXFORD
UNIVERSITY PRESS

Great Clarendon Street, Oxford, OX2 6DP,
United Kingdom

Oxford University Press is a department of the University of Oxford.
It furthers the University's objective of excellence in research, scholarship,
and education by publishing worldwide. Oxford is a registered trade mark of
Oxford University Press in the UK and in certain other countries

© François Louchet 2021

The moral rights of the author have been asserted

First Edition published in 2021

All rights reserved. No part of this publication may be reproduced, stored in
a retrieval system, or transmitted, in any form or by any means, without the
prior permission in writing of Oxford University Press, or as expressly permitted
by law, by licence or under terms agreed with the appropriate reprographics
rights organization. Enquiries concerning reproduction outside the scope of the
above should be sent to the Rights Department, Oxford University Press, at the
address above

You must not circulate this work in any other form
and you must impose this same condition on any acquirer

Published in the United States of America by Oxford University Press
198 Madison Avenue, New York, NY 10016, United States of America

British Library Cataloguing in Publication Data
Data available

Library of Congress Control Number: 2020939783

ISBN 978–0–19–886693–0

DOI: 10.1093/oso/9780198866930.001.0001

Links to third party websites are provided by Oxford in good faith and
for information only. Oxford disclaims any responsibility for the materials
contained in any third party website referenced in this work.

Use At Your Own Risk

Mountaineering, and more particularly off-piste skiing, are dangerous and can involve exposure to avalanches and other hazards.

The present book aims at disclosing and disseminating the essence of avalanche triggering processes. The resulting guidelines cannot eliminate these hazards, but they can help understand and manage them.

The content provided in this book is given "as is" and in no event shall the author be liable for any damages, including without limitation damages resulting from discomfort, injury or death, claims by third parties or for other similar costs, or any special incidental or consequential damages arising out of the use of this publication.

This book is obviously no substitute to training, experience, skill and wisdom.

Contents

Foreword and Acknowledgements		ix
1	Introduction	1
2	Snow, an Intriguing, Complex, and Changeable Solid	5
	2.1 From ice to snow	5
	2.2 Snow crystals	6
	2.3 From snowfalls to snow layers	8
	2.4 Snow as a granular medium	10
	2.5 Snow as a porous medium: the concept of percolation	11
3	Deformation, Fracture, and Friction Processes	14
	3.1 Deformation of solids	14
	3.1.1 Elasticity	14
	3.1.2 Plasticity	15
	3.1.3 Static vs dynamic loadings	16
	3.2 Fracture initiation and extension	16
	3.3 Griffith's criterion	17
	3.4 The brittle to ductile transition	20
	3.5 Coulomb's law of friction	22
4	Slab Avalanche Release: Data and Field Experiments	25
	4.1 Geometry and dynamical characteristics	25
	4.2 Statistical aspects: scale invariance	27
	4.3 The weak layer, starting point for slab avalanche release	29
	4.3.1 Propagation Saw Tests (PST)	30
	4.3.2 Sintering (clotting) experiments	31
	4.4 Stability and bridging indexes	32
5	Slab Avalanche Modeling	36
	5.1 Old myths and beliefs to shoot down	36
	5.2 Basis for modeling	37
	5.3 Statistical approach: Playing with cellular automata	38
	5.4 Sliding or sticking?	42
	5.5 Slab avalanche release in four steps	43
6	Superficial and Full-Depth Avalanches	45
	6.1 Loose snow avalanches	45
	6.2 Full depth avalanches	47
	6.2.1 Observations	48
	6.2.2 Different kinds of snow involved in full-depth avalanches	49
	6.2.2.1 Dry dense snow	50

viii   *Contents*

		6.2.2.2	Wet snow	50
		6.2.2.3	Soggy snow	50
	6.2.3	Release mechanisms		51
		6.2.3.1	Basal crack destabilization	51
		6.2.3.2	Wet avalanche release	52
		6.2.3.3	Soggy snow avalanche release	55
		6.2.3.4	Dry snow full-depth avalanches	55
	6.2.4	Propagation and arrest		55
6.3	Summary			56

7 Snow and Avalanches in a Climate Warming Context 57

   7.1 Climate change 57
   7.2 Possible consequences for avalanching 59

8 Summary and Conclusion 61

Appendix A Complexity and Critical Phenomena 63

   A.1 From simple to complex systems 63
   A.2 Scale invariance and self-organized criticality 65

Appendix B Modeling a Fluid to Solid Phase Transition in Snow Weak Layers: Application to Slab Avalanche Release 69

   B.1 A fluid to solid phase transition in healable granular materials 69
   B.2 Application to slab avalanche release 72

Appendix C Stability of a Sintered Weak Layer Disk Surrounded by a Ring-Shaped Fluid Weak Layer Zone 77

References 79

Index 83

Foreword and Acknowledgements

Do not go where the path may lead, go instead where there is no path and leave a trail
Ralph Waldo Emerson

This was a magnificent, resplendent, brilliant, gorgeous winter Sunday morning. After a week-long period of snowstorms, an incredible sun was shining on tremendous amounts of freshly fallen snow. I was reluctantly wearing a pair of terrible snowshoes instead of my preferred cross-country skis, and we were walking with my wife Marie and a couple of friends, Jacqueline and Serge Macel, between the Valloire ski resort and "lac des Cerces". A short distance beyond the small shepherd cabin of Plan Lachat, at the foot of the Galibier pass, we suddenly heard a loud bang coming from the Pointe du Vallon, and immediately noticed a huge avalanche tumbling down from the very top of the steep slope, readily turning into an airborne powder flow. "Watch out, watch out!" Serge shouted, pointing at two skiers a significant distance below, trying to escape the avalanche front. They obviously could not succeed, and were readily swept out by the impressive flow like miserable tiny beetles. The avalanche went on, crossed the small Valloirette creek, climbed up the opposite riverside in our direction, and stopped at about 100 m from where we were standing. The whole thing got rooted to the ground. Not the slightest noise. Not the slightest sign of life. I hurtled down the path towards Valloire trying to get help, yelling to other trekkers down-slope, who in turn tried passing on the message to other hikers, while my companions were desperately scanning the closest part of the avalanche flow. After a while, I was flown over by a helicopter. I clearly heard it land higher up, out of my sight, and take off again only after a few seconds. I received the explanation 15 minutes later while rushing back to join my group: I met two guys skiing down towards Valloire and asked them what was going on up there. They furiously replied that they were themselves the skiers caught by the avalanche, and that I shouldn't have given the alarm or called the helicopter! I was told later on by my group that the head of one of the skiers was jutting out above the snow level. He succeeded in getting out, found his companion, and extracted him from the avalanche. The latter was terribly angry to have lost…his glasses! And his main comment was that he was a professional mountain guide, and that they were not responsible for the avalanche release, for the "obvious" reasons that the avalanche started at a significant distance above them, and that they have already crossed this slope several times in the past without any problem.

As far as I remember, that was in February 1997. At this time, I didn't know much about avalanches, but I was deeply convinced that such statements were stupid. All I had to do was figure out why. This is how I started my research work on avalanches.

This work could not have been achieved and the present book would not have been written without sometimes fortuitous, but always invaluable and priceless encounters, often turning into warm, long lasting and unforgettable friendships. I will especially mention and gratefully acknowledge:

Malcolm Heggie and Jany Thibault. Both of them sadly passed away recently. Both of them had been working on dislocation core structures in covalent crystals, Malcolm essentially on theoretical models in diamond and ice, Jany and myself on electron microscopy and modeling dislocation cores in semiconductors. Dislocations in such crystals and in ice are quite similar. In addition to my immoderate taste for mountaineering, this was probably one of the reasons for my interest in ice and snow. I first met Malcolm in the seventies. We visited each other several times in the universities of Toulouse, Exeter, and Grenoble. He introduced me to the concept of soliton, in the sense of a solitary broken bond in a covalent dislocation core, on which I built theoretical models of dislocation mobility in pure and compound semiconductors. Malcolm and I shared numerous conferences, particularly with Jany during several two-week workshops in Aussois. Jany was also a brilliant physicist, and a top-range mountaineer. She used to organize a few ski and trek programs during such conferences. As for Malcolm, after having tried to teach him cross-country skiing in thick and steep forest environments, I remember him complaining his skis were longer than the average tree separation.

Kolumban Hutter, met during my very first conference on avalanches at Innsbruck EGS symposium in 2000. I started my talk saying something like: "*I am a physicist. I've been working on plasticity and rupture of crystalline materials for more than 20 years, but I do not belong to your scientific community. This is my first contribution to avalanche research*". I had no idea what would be the reactions of the audience to my presentation. Kolumban was the very respected session chairman and, as such, was sitting in the front row, just facing me. I was presenting a fairly simple but original slab avalanche model, showing that a transition from a tensile crown crack instability to a shear basal one occurred for a universal angle $A\cos\sqrt{2/3} = 35.3°$. The chairman's comment was: "*Let's define that as Louchet's angle!*" Thank you, Kolumban, for having immediately trusted me during my avalanche initiatory rite!

Jérôme Weiss and Paul Duval, with whom I have been working on ice plasticity and fracture at the Grenoble Glaciology Department for more than 10 years. Thank you Paul for your invaluable glaciologist experience during our work on ice slip geometry, dislocation cross-slip, and dynamic recrystallization. Thank you also Jérôme for your pioneering work on dislocation avalanches by acoustic emission analysis, which was the starting point of a long and fruitful collaboration on self-organized critical dislocation dynamics. With our common PhD student Thiebaud Richeton, we showed in particular that such a behavior was shared by a number of other materials, and we evidenced a quite general "mild" (Gaussian) to "wild" (scale-invariant) transition controlled respectively by short-range or by long-range elastic interactions. Based on our numerous discussions, this period also gave me the opportunity of revisiting Hall–Petch law and Andrade creep in terms of approaches of a critical point.

Jérôme Faillettaz, a young mechanical engineer fond of skiing and mountaineering. I was his PhD thesis supervisor. During innumerable coffee-fueled discussions, we discovered together unknown and unexpected fantasies of avalanches. We scoured a number of international meetings, carefully avoiding awfully clean and expensive conference hotels, sharing a can of beans in a cheap room in Davos, or spending a couple of nights in a small tent in the hills during an EGS conference in Nice. Jérôme is now a Senior Researcher at Zürich University. Thank you so much Jérôme for your disrespect to scientific or any other type of "authority", your enthusiasm, and your confidence. Thank you also for your careful reading of the proofs of this book, and your quite sensible comments.

Jean-Robert Grasso, who helped us enter the universe of Self Organized Criticality. Together with Jérôme, we designed a specific cellular automaton, disclosing the mechanisms responsible for such a scale-invariant (and probably self-organized) critical behavior, and merging all gravitational flows into a single formalism. Thank you Jean-Robert for your brilliant idea to spend a year in UCLA, while Jérôme and myself were in Grenoble. Owing to the time lag between Grenoble and Los Angeles, and the back and forth daily mailing, the succession of our respective sleeping and active periods resulted in a quasi-continuous working time. Our paper on the cellular automaton in *Physical Review Letters* was written within an incredible short time, and qualified for illustrating the front cover of the journal.

Alain Duclos, met somewhere in the Canadian Rockies. We were attending a snow and avalanche workshop in Penticton (British Columbia), a small town with a nice old wooden shop serving mugs of coffee and beer, and selling tons of second-hand books. I bought a number of them (books, not mugs), much to my backpack's and my own backbone's misfortune. Alain was quite interested in the work presented at the meeting by Jérôme and myself. We decided to continue our passionate discussions when coming back home. We did. I became the secretary of his newly created Data-Avalanche Association, now widely recognized in the avalanche field. Alain and many other members of the Data-Avalanche association taught me much of their long-standing field experience. We shared numerous discussions, working sessions, field experiments, and also friendly working dinners. A lot of thanks to all of them. Particular thanks again to you Alain for sharing your invaluable field knowledge and analysis during proof reading of this book, and also to Céline Lorentz for designing a schematic in chapter 5 of this book.

Joachim Heierli, who proposed the innovative "anticrack" concept. Thank you Joachim for your participation in field experiments in Aussois, and for sharing spirited scientific discussions in Edinburgh, Davos, Grenoble, and Freiburg.

My brother Jean for his careful and detailed reading of the manuscript, and his meaningful remarks.

My four anonymous referees, for their interesting and valuable comments, that helped me improve and complement several parts of the book, and also my Editors for the particularly efficient and trustworthy relationship I had with them.

Finally, and above all, I would like to pay a special tribute to my wife Marie, to whom I dedicate this book. She has been enduring and still endures the formidable and terrific role of a researcher's wife, living every day with a guy able to suddenly stop in front of any weird object by the roadside, immediately falling into a contemplative ecstasy, keeping his brain in a state of intense excitement until the deepest and ultimate secrets of the "thing" have eventually been revealed. She deserves my profound admiration for her patience, and my infinite gratitude for her constant and unfailing support.

1
Introduction

A journey of a thousand miles begins with a single step
Lao-Tzu

The names of many places and villages in European mountains refer to memories of old, more recent, or recurrent avalanching events, as "lavancher" or "lavachet" in local French-speaking alpine dialects, coming from "labina" in Latin. This is also the root of the term "Lawine" in German, "avalanche" in English and French, "avalancha" in Spanish, and "valangha" in Italian.

An avalanche may be defined as the destabilization and flow of part of the snow cover. We shall essentially deal here with the former, focusing on avalanche triggering mechanisms. Studies of avalanche flow processes indeed mainly involve fluid mechanics, and are usually described by classical Navier–Stokes equations, despite the difficulty due to heterogeneity (e.g. vertical snow density gradient) and specific non-linear aspects of dynamical snow behavior. This is another and quite interesting story to look at, but partly out of the main scope of the present work. Meantime, avalanche triggering mechanisms have been debated for decades, and need some re-foundation on clear scientific bases. This is the main goal of the present book.

Snow avalanches share a number of characteristics with some other types of gravitational flows. Avalanches shown in Figs 1.1 and 1.2 perfectly illustrate slab avalanches and loose "snow" avalanches, as defined hereafter, except that they actually are gypsum sand and not snow avalanches! Another feature shared by gravitational flows is given in chapter 5, showing common statistical characteristics between snow avalanches, landslides, and rock-falls, all of them belonging to the same class of critical phenomena.

Avalanche studies are at the intersection of several traditional fields of science or practice. Each community (physics, mechanics, mathematics, practitioners, etc.) has its own language, which may yield some misunderstandings. The present book is written in a language used in physics, but equivalents will be given for clarity if necessary. In this respect, it should be useful to make clear a few definitions and several idioms that are used to characterize avalanches.

The snow cover is a layered structure built up during successive snowfalls. Upper layers may be destabilized and glide down as a whole, resulting in "slab avalanches". The glide plane, the interface between the gliding slab and the older snow substrate, is known as the "basal plane". It initially consists of a "weak layer" (*WL*) made of brittle snow, whose collapse may trigger the avalanching process. But the snow cover may also glide as a whole on the underlying bare ground, giving rise to so-called "full-depth" avalanches.

More precisely, three main types of snow avalanches are usually distinguished:

i) Slab avalanches. Their release results from the initial failure of an underlying weak layer (*WL*) that separates two adjacent snow layers. Such a failure, that may or may not result in avalanche release, usually propagates beneath the slab over distances ranging from meters

Fig. 1.1 *Gypsum sand slab avalanche, illustrating slab snow avalanches, White Sands National Park, New Mexico (USA). (Photograph by François Louchet).*

to kilometers, and is associated with a downward displacement (or "settlement") of the slab, and sometimes with an audible "whumpf", an onomatopoeic term for the muffled noise produced by the settlement. They are responsible for most human fatalities, and therefore deserve specific interest.

ii) <u>Loose snow avalanches</u>. In cold fluffy snow, i.e. with low cohesion, they occur preferentially on steep slopes. They are often thought to be triggered by the destabilization of a few snow grains that knock out a couple of other ones, and so on, usually resulting in a narrow snow slide flowing down as a superficial combination of tiny sluffs from a quasi-punctual starting point (Fig. 1.2), and gradually growing in size. Their bed surface is ill defined. They may be quite harmful when pouring down into narrow gullies. By contrast, in wet snow conditions, low cohesion results from lubrication by melt water, and may help release of loose snow avalanches on gentler slopes, with significantly larger sizes. A detailed description of such mechanisms will be found in (Daffern 1992, Tremper 2008).

iii) <u>Full-depth avalanches</u>, actually encompassing all other types together, which is the reason why a variety of full-depth avalanche definitions are found in the literature.

A few other terminologies are often used:

i) The distinction between "<u>dense</u>" and "airborne <u>powder</u>" avalanches essentially refers to avalanche flow processes. As we shall focus on triggering mechanisms, this question has in principle no call to be discussed in detail here. However, some avalanche flow characteristics

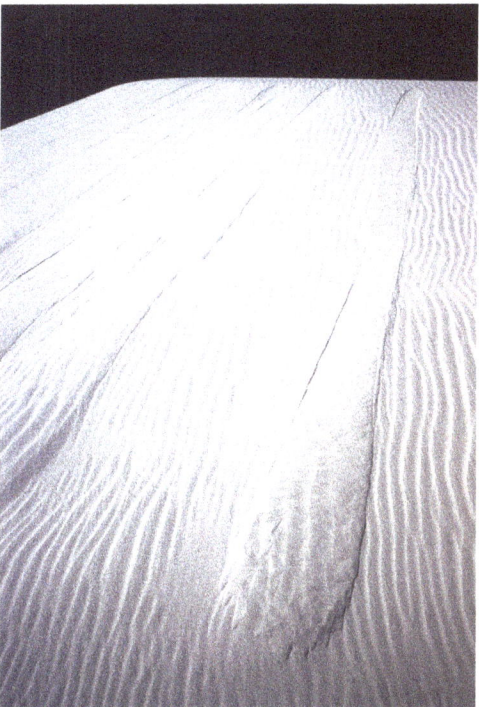

Fig. 1.2 *Loose dry gypsum sand avalanches, illustrating loose snow avalanches, White Sands National Park, New Mexico (USA). (Photograph by François Louchet).*

  may be determined by initial conditions, and powder flows may originate from particular triggering processes.

ii) Slab avalanches can be <u>artificial</u> (or accidental) or <u>spontaneous</u> (or "natural"), depending on whether they result from some external action (skier, animal, explosives, etc.) or not. Spontaneous avalanches necessarily involve some time-dependent ingredients, as an overload due to a cornice rupture, a new snowfall, or accelerating creep (i.e. viscous flow) that may bring the system from ductile deformation to brittle failure (Gubler and Bader 1988). However, the possible role of the *WL* is still under debate in this last case.

Owing to the complexity and variability of the snow cover and of unexpected variations of weather conditions, forecasting avalanche release is a formidable task. Our belief is that an increased knowledge and understanding of underlying mechanical and thermodynamical processes can significantly help both hazard control and mitigation measures.

The main goal of the present book is thus to disclose and analyze the main processes involved in avalanche release. It seems useful to start in chapter 2 with a short review of the basics of snow structure and topology. Snow being a complex arrangement of ice crystals, themselves found in oodles of geometrical shapes, sizes, and formation mechanisms, we shall essentially focus on those that are more directly involved in avalanche release. Readers interested in further developments are referred to specialized textbooks.

We shall essentially focus on two snow peculiarities. Since the snow cover results from an accumulation of snowflakes, it may be considered as a granular material, with quite original properties due to the unusually large grain surface vs volume ratio, and to their changeable healing propensity. Snow also being a mixture of ice, air, and water, the topological concept of percolation is of interest to deal with stress distribution in the snow cover, and will be briefly discussed.

Chapter 3 will be dedicated to some mechanical and physical concepts ruling deformation, fracture, and friction processes, with particular attention paid to the simplicity of the analysis, but without betraying the scientific validity of the arguments.

Chapters 4 and 5 will get into the very heart of the matter, with a thorough exploration of slab avalanche release mechanisms. First, observations and field experiments will be analyzed. The modeling section will empanel digital simulations and analytical approaches, whose results will be extensively discussed.

Chapter 6 will deal with superficial and full-depth avalanche triggering, discussing the various possible types and corresponding mechanisms, essentially in terms of self-organized criticality for the former, and of percolation for the latter.

Finally, chapter 7 will tentatively discuss the expected influence of the present and unprecedented climate warming on avalanching activity and associated hazards.

The reader is also encouraged to visit the Data-Avalanche site http://www.data-avalanche.org/ for further and more practical information, more particularly on risk management and mitigation strategies.

2
Snow, an Intriguing, Complex, and Changeable Solid

> *We must always tell what we see, but above all, and this is more difficult, we must always see what we see*
>
> Charles Péguy

The purpose of the present chapter is to give the minimum basic concepts that will be useful for understanding avalanche problems. For more information, the reader is referred to the considerable bunch of snow treatises available in libraries and bookshops, and more particularly to chapter 3 of the excellent book by Tony Daffern (1992) on avalanche safety. It is however worth noting that the very particular structure of snow, made of complex and changing mixtures of ice, air and water, endows this material with a considerable variety of physical properties. The topological concept of percolation, that can be conveniently used for exploring these properties, will be discussed at the end of the chapter.

2.1 From ice to snow

Figure 2.1 shows the water phase diagram. Line AD separates solid (left) from liquid (right), and AE separates liquid (top) from vapor (bottom). As shown by the horizontal line BC, under usual pressure conditions found on the Earth's surface (so-called "normal conditions"), water solidifies at 0°C (into 1H type ice crystals, with a hexagonal crystallographic structure), and boils at 100°C.

Beyond the "critical point" E, liquid and vapor phases cannot be distinguished any more. There is no sharp transition in this case between these phases. In other words, phase boundaries vanish. We deal in this case with a "supercritical" fluid. Actually, the concept of criticality is by far more general than its application to the water phase diagram. We shall return to this concept in Appendix A for other applications.

A particularly interesting and well known feature of ice crystals is illustrated by the negative slope of line AD: starting from a state in the solid phase, a pressure increase at constant temperature brings the system through the AD line up to the liquid state. The physical reason for this counter-intuitive peculiarity is that, due to the specific molecular bonding of 1H crystals, ice density is lower than that of water, i.e. ice takes up a larger volume than the same weight of water. This property results in floating ice cubes, or at a different scale, drifting icebergs. Taking the problem the other way round, another and obvious consequence of this specificity is that trying to shrink the volume of a piece of ice by external pressure favors melting, since liquid water satisfies with a smaller volume than solid ice.

Snow being made of ice, it inherits the same property, which has interesting consequences in ski practice: in "warm" snow, ski pressure favors the formation by local melting of a thin water

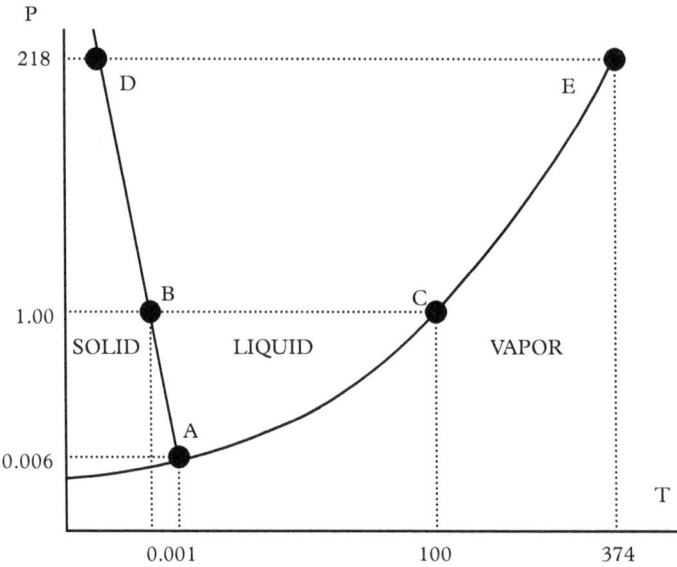

Fig. 2.1 *Phase diagram of water. Pressure is in atmospheres, and temperature in °C. A is the triple point where the three phases (solid, liquid, vapor) coexist. E is the critical point, beyond which there is no sharp transition between liquid and vapor. The negative slope of the DA line is responsible for quite specific characteristics of the water solid/liquid transition (see text).*

layer that helps sliding, despite the fact that warming by ski friction on snow also contributes to the same melting effect. This is by far less efficient in colder snows, a phenomenon well known by cross-country skiers, who have the strange feeling of skiing on dry sand instead of snow, more especially if they are using the skating technique.

2.2 Snow crystals

Snow crystals nucleate from a supersaturated atmosphere, helped by dust particles through a reduction of interfacial energy. They nucleate as single crystals, i.e. in which water molecules are precisely arranged along parallel directions and stacked planes. The well-defined V-shaped geometry of water molecules (Fig. 2.2) results in precise molecular arrangements when condensed in the solid state. In the 1H solid phase (our well known common ice, stable under our Earth's pressure and temperature conditions), the 104.45° angle between O–H bonds in the free molecule slightly deforms up to 109°, in order to comply with a stable hexagonal crystallographic symmetry (Schulson and Duval 2009). By contrast with Oxygen atoms, Hydrogen ones exhibit a limited long-range ordering, obeying Bernal–Fowler rules (Pauling 1933):

i) two Hydrogens are located close to each Oxygen.
ii) each O–O bond must not contain more than one Hydrogen.

However, such features are of little importance in snow mechanical properties.

Due to the 109° angle of the H₂O molecule, snow crystals exhibit a simple 6-fold symmetry, resulting however in impressive oodles of complicated shapes, depending on temperature, temperature gradients, and humidity during nucleation and growth. A few examples are shown in Figs 2.3 and 2.4. As a detailed description of such crystals is beyond the scope of the present work, the reader is referred to (Daffern 1992, Libbreght 1999).

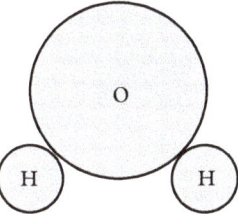

Fig. 2.2 *Schematic model of a water molecule H_2O. The angle between O–H directions in the free molecule is 104.45°.*

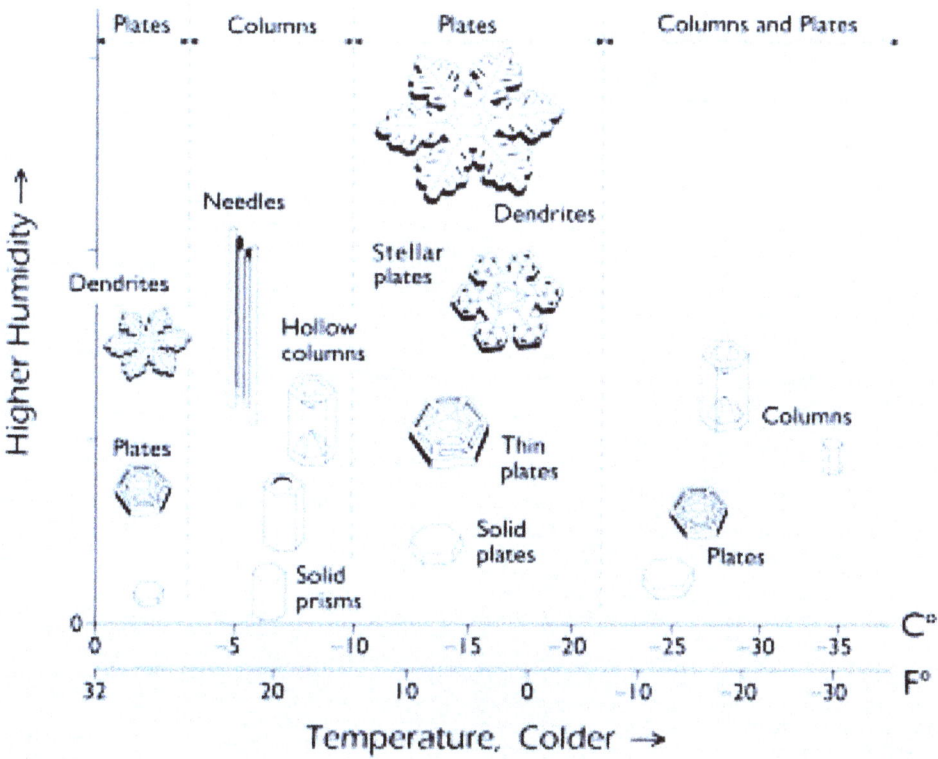

Fig. 2.3 *Snow crystal morphology vs temperature and humidity. (After Ken Libbrecht, kgl@caltech.edu, http://www.snowcrystals.com/science/science.html).*

Fig. 2.4 *Examples of snow crystals: top left and right, stellar dendrites (i.e. star-shaped tree-like crystals); bottom left, fernlike stellar dendrites (containing many side branches); bottom right, rimed crystals, resulting from collisions between snow crystals and freezing water droplets. (After Ken Libbrecht, http://www.its.caltech.edu/~atomic/snowcrystals/class/class-old.htm).*

After nucleation, such crystals aggregate during their motion in the atmosphere into snowflakes, each of them being made of a combination of single crystals of various orientations. Due to the intricate shapes of snow crystals, their random aggregation into snowflakes incorporates a significant amount of air, which results in a low density. This is why snowflakes fly around during snowfalls, in strong contrast with hail showers. This is also why they can be easily transported by wind, which may result in so-called wind slab formation.

The white color of snow differs from the transparent and slightly bluish aspect of bulk ice, due to the huge surface vs volume ratio, that favors light scattering.

2.3 From snowfalls to snow layers

Snow falls on the ground forming successive layers, whose structure and properties may evolve with time, helped again by humidity, temperature, and temperature gradient conditions, and various types of loading (further snowfalls, skiers, snowmobiles, grooming machines, etc.). In some cases, the top layer may be formed by wind transportation, forming improperly named "wind slabs".

The cohesion between such layers also evolves with time. Failure of one of these interfaces, as mentioned in the introduction, may destabilize as a whole the layer stacking sequence located above the destabilized interface, resulting in an incipient slab avalanche. As a consequence, the term "slab" refers to the group of such destabilized stacked layers, which cannot be properly defined until triggering occurs. Such slabs are responsible for the majority of avalanche fatalities. This is why two chapters (4 and 5) will be most entirely dedicated to slab avalanche release.

Other types of ice crystals may form in the stacking sequence of the snow cover, at the surface or at interfaces of already deposited layers. This is the case for instance of surface hoar, facets, and depth hoar. Surface hoar is made of those superb ice crystals that grow at snow surface in "warm" and humid atmospheres, due to the significant temperature gradient at the surface of a colder snow cover during cold and clear nights. These well-known shiny flakes provide incredibly smooth ski sliding. If they are buried during a snowfall before transformation into stronger structures, they become a layer (still named "surface hoar"!) on which the slab may potentially slide down very easily.

Facets and depth hoar also grow under thermal gradients, often (but not always) at interfaces between different snow layers: indeed, during clear nights, the external temperature goes down, whereas deeper snow layers keep warmer, due to the thermal flux from the ground (geothermal flux). It is usually thought that this phenomenon takes place on north slopes, as the external temperature is colder than on south ones, at least in the northern hemisphere (readers from the southern hemisphere should translate this sentence the other way round!). It is worth noting however that this mechanism may also occur on other slope orientations, particularly in early winter during which the atmosphere cools down due to a weak and low sun, whereas the ground keeps warmer. Under such a temperature gradient, the system is out of equilibrium, water molecules evaporate from the warmer bottom layer ("sublimation"), and condense on the colder bottom part of the upper layer, resulting in a lace of delicate brittle crystals.

In all cases, resulting intermediate interfaces consist of (or transform into) granular aggregates of polyhedral ice grains bonded by brittle ice bridges, usually known as "weak layers" (*WL*) (Fig. 2.5).

As the metamorphic transformation occurs at constant volume, the average density of the *WL* is comparable to those of top and bottom layers. Yet, it may exhibit a larger brittleness for at least two reasons:

i) It is made of bigger crystals separated by larger flaws. Griffith's criterion developed in chapter 3 states that larger flaws correspond to a reduced toughness.

ii) Despite the fact that the *WL* average density is similar to that of top and bottom layers, it may vary across the *WL* thickness, the lighter zones being significantly more brittle than the denser ones.

In addition, the lower thermal conductivity of the lighter layer increases the local temperature gradient, enhancing the metamorphic transformation rate. For a similar reason, *WLs* may also form on both sides of freezing crusts (this is called a "super gradient").

For all these reasons, *WLs* are recognized to play a key role in snow slab avalanche triggering processes. It is therefore of interest to understand in more details the weak layer behavior, in order to be able to predict which conditions may favor slab avalanching, as detailed in chapters 4 and 5.

Fig. 2.5 *Typical weak layer section. Bonneval sur Arc (Savoie) 22 February 2017. Scale is given by 2 cm × 2 cm squares. (Photograph by Alain Duclos).*

2.4 Snow as a granular medium

Granular matter is made of assemblies of solid grains, whose sizes may range from nanoscopic up to macroscopic scales. Properties of such ensembles strongly depend on grain sizes, through the volume vs surface ratio $R=V/S$. As volumes and surfaces scale respectively as the 3rd and 2nd powers of grain size, R is homogeneous to a length. As a consequence, the behavior of granular ensembles is essentially ruled by volume properties in the case of bigger grains (large R values), and by surface in the case of smaller ones.

In this respect, granular media may be defined as a wide intermediate stage in which both volume and surface effects have to be simultaneously taken into account to understand physical properties.

In the case of mechanical (and more specifically dynamical) properties for instance, granular media behavior is ruled by the balance between grain weight and inertia, which are volume parameters on the one hand, and contact interactions, obviously of surface nature on the other hand.

In the limiting case of powders, small R values enhance surface effects during contacts, as for instance friction, cohesion, or chemical reactivity (including water capillary effects for sand or snow for instance). In the opposite case of large R values, inertial effects dominate, resulting in strong changes in bulk structure during collisions. This is the case for shocks between icebergs, which may yield bulk damage and fracture. In comparison, collisions between small flying beetles (with small R values) are much less damaging than those between heavy birds or cars, whose external skin or shell cannot resist inertial effects.

The lower bound of this domain is usually considered to be at around 1 µm. Above this size, indeed, grains are large enough to be unaffected by Brownian motion (i.e. thermal fluctuations), which means that surface effects will not overwhelm volume effects any more. The upper bound of the granular media domain is more arbitrary and difficult to define precisely, and depends on investigated physical properties, but the centimeter or decimeter range looks reasonable.

Granular media may be found under two main states, as deformable and possibly flowing condensed matter when relative velocities between grains are low enough, as for full depth avalanches for instance (chapter 6), or as suspensions in a fluid (liquid or gas) in the opposite case (e.g. airborne powder avalanches, or wind transportation of snow). In the condensed state, stresses are not homogeneously distributed as in a "traditional compact" solid, but are concentrated along particular patterns called "force chains". Such chains may act as screening shields for other grains embedded in between, which experience lower stresses. Formation of arches, that may support the weight of grains located above, and protect those lying underneath, is a particular case of force chains.

Snow is a very particular case of granular materials, due to the complicated "hairy" surface of snow grains, often made of elongated arms (dendrites) as shown above (Figs 2.3 and 2.4), and characterized by huge specific areas. For comparable grain sizes, this geometrical specificity shifts R values towards much smaller scales as compared to more classical granular media made of spherical, polyhedral, or more generally convex shaped grains. Such a peculiarity has various consequences:

i) Due to smaller R values at comparable grain sizes, phenomena related to grain contacts are enhanced. This is the case for friction that hinders sliding and associated shearing, for increased cohesion, and also for grain welding that has fundamental consequences on avalanche release and on flow arrest processes as well (chapters 5 and 6, appendix B).

ii) Snow grain arms are particularly brittle; snow grains may easily collapse under stress; this is the case for weak layers (Fig. 2.5, and chapters 4 and 5), resulting in a significant increase of R values, becoming more comparable to classical granular matter, and making slide and shear phenomena much easier.

iii) Under most physical conditions found on Earth, snow is fairly close to its melting point. Due to water molecules' diffusion on surfaces, driven by surface tension reduction, the intricate shapes of snow crystals may also evolve into more rounded ones, with larger R values, leading to the same consequences as in ii).

Snow is thus a very particular granular medium. When made of cold, loose, and dry crystals, it may be indeed dealt with as a granular solid, or as a granular fluid in the case of wind transportation for instance, but cohesive snow is more conveniently described as a porous medium, as discussed hereafter.

2.5 Snow as a porous medium: the concept of percolation

The porous character of snow is indeed quite important, as it may affect mechanical properties. By contrast with bulk ice, snow may be considered as an ice foam that may contain a significant amount of air and water. It may for instance collapse under stress into a denser material, or sinter under favorable temperature conditions. As a consequence, it may exhibit changing physical or mechanical properties, and usual laws ruling bulk solids may not directly apply. This is the case for instance for Coulomb's friction law, as shown and analyzed in chapter 3.

Other properties are controlled by snow topology, particularly in the case of full depth avalanches discussed in chapter 6. The topological concept of percolation, reminiscent of coffee making, is quite important in various physical problems, and particularly in yielding and fracture mechanisms. It can be illustrated in a very simple way, as follows.

Let us consider a coffee filter filled with coffee powder, and let us pour water on it. Three different situations may be considered:

i) If the powder is firmly crammed down into the filter, hot water poured on it would stay at the surface, unable to go through the powder. In this case, it is possible to find connected paths made of coffee powder grains in mutual contact, resulting in "force chains" that go through the entire system. Coffee grains are said to "percolate" through the system, but water does not, and coffee making becomes a complicated task.

ii) If the coffee powder is less tightly packed, water would be able to find a way through the powder, down to the coffee pot. It percolates through the system, but the coffee powder still percolates, which prevents its mechanical collapse. This is called "bi-percolation".

iii) if too much water is poured at the same time on the coffee powder, the mixture turns into a fluid in which coffee powder grains float freely. Water percolates, but coffee grains do not any more. This is the situation found in "turkish coffee" making.

Physical, and more specifically mechanical, properties of random media drastically change at so-called percolation thresholds, where isolated clusters become connected into a theoretically infinite network, or conversely.

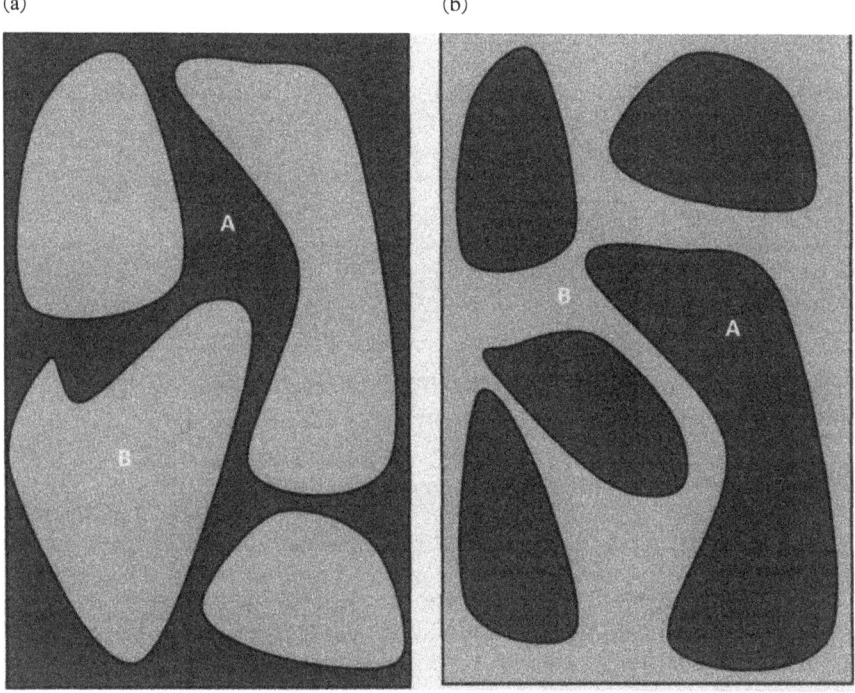

Fig. 2.6 *Percolation: (a) A percolates in B; (b) B percolates in (A). A and B can mutually percolate only in 3 dimensions (bi-percolation).*

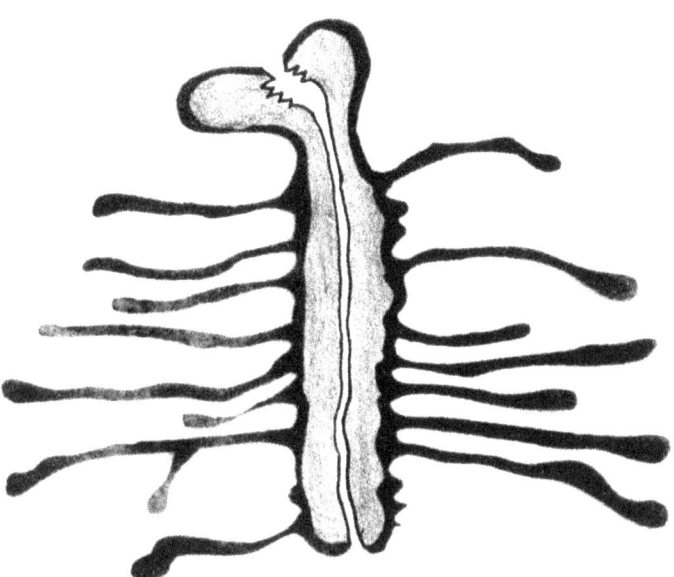

Fig. 2.7 *A two-dimensional beetle necessarily split into two separate parts due to its digestive system. (Adaptation by François Louchet of an artwork by Martine Rey).*

We shall see in chapter 6 that, in a similar way as for coffee, rain water may percolate through snow layers, and how bi-percolation or "mono-percolation" may result in quite different wet snow avalanche types.

It is worth noting that bi-percolation would be impossible in a two-dimensional space. For instance, in a 3-d coffee maker, water is able to use the 3rd dimension to bypass a continuous chain of powder coffee grains, which is impossible in 2-d. This is a privilege of three-dimensional spaces. We should be delighted to realize that if we were two-dimensional beings, we would not be allowed to enjoy coffee drinking, to say nothing of the fact that, for the same reason, the whole ingestion-digestion process would split the drinker into two separate parts (Fig. 2.7)! Such a situation may be encountered in avalanche glide surfaces (that are 2-d objects), A and B being for instance snow and water for full-depth avalanches sliding on bare ground, or collapsed and non-collapsed zones in slab avalanche weak layers.

3
Deformation, Fracture, and Friction Processes

*What gets us into trouble is not what we don't know.
It's what we know for sure that just ain't so*

Mark Twain

In order to avoid misunderstandings in the discussion of avalanche release mechanisms, precise definitions of involved physical (and more particularly mechanical) quantities are required. Usual mechanical properties of solids are defined hereafter on this basis.

3.1 Deformation of solids

Stress is defined as the force per unit surface, expressed in Pa, kPa (10^3 Pa), or MPa (10^6 Pa), in the same way as pressure, which is a particular case. Stresses may indeed describe different loading geometries. Uniaxial tensile stresses are responsible for lengthening rubber bands, uniaxial compressive ones for shortening car suspension coils, shear components are involved in resistance to ski glide, etc. For this reason, stresses are usually expressed in the form of "tensors", mathematical objects written as 3 × 3 tables, containing nine "scalar" stress components. In avalanche problems, the most useful components are tension, compression, and shear. For the sake of simplicity, they will be referred to as scalars, expressed in Pa, kPa, or MPa.

Strain is defined as relative deformation, i.e. for instance the lengthening of a sample divided by its length. It is dimensionless (i.e. has no units), but is often expressed in %. Strictly speaking, in the same way as stresses, strains are tensors. Their different components may be distinguished, but we shall use them separately as tension, compression, or shear strains. Stress-strain curves, which characterize deformation properties of solids, exhibit several stages, as discussed now.

3.1.1 Elasticity

A typical stress-strain curve of a solid is shown in Fig. 3.1. It usually starts with a so-called elastic domain in which strain is proportional to stress. In this domain, strain is reversible, which means that the original shape is restored upon stress release.

Elastic properties are characterized by various quantities, depending on the involved stress tensor components. They are usually referred to under the generic name of "***stiffness***":

Young's modulus (E) describes the elastic response of a solid experiencing a load along a uniaxial compression or tensile axis. It is defined as the stress to strain ratio, and represented by

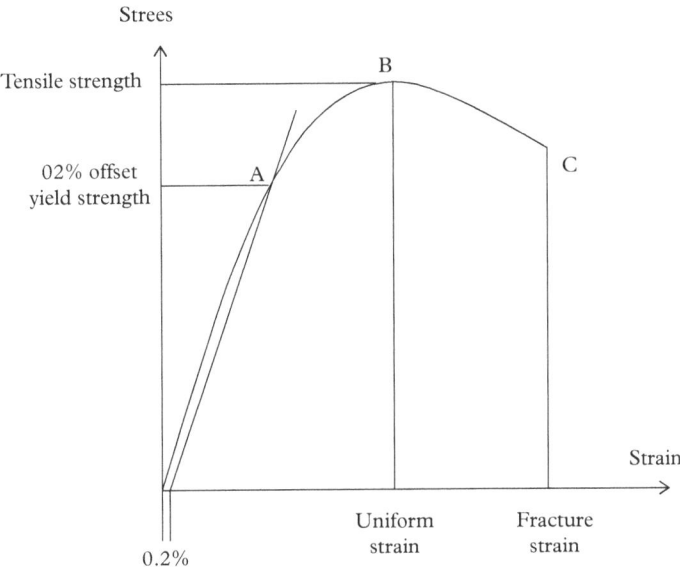

Fig. 3.1 *Schematic stress-strain curve, showing the elastic and plastic domains, and definitions of various mechanical quantities. The curve represents the stress vs strain variations under constant strain rate. The yield stress is usually defined as the point A where the stress-strain curve departs from the elastic domain straight line by an amount of 0.2%. Beyond this point the material deforms plastically, i.e. it does not recover its initial shape upon stress release. The so-called tensile strength corresponds to the maximum stress (B), and is the end of the homogeneous plastic deformation stage. Beyond B, the flow stress decreases due to strain localization, ending up with ductile rupture at C.*

the slope of the stress-strain curve in the elastic domain (Fig. 3.1). Strain being dimensionless, E is measured in Pa, kPa, or MPa. It is also referred to as the **elastic modulus**.

The **shear modulus** (G) describes the solid response to a shear stress (instead of a tensile stress), and is defined as the ratio of shear stress to shear strain. It is also measured in Pa.

In compact solids, both uniaxial and shear elastic deformations take place at constant volume.

The **bulk modulus** (K) describes volumetric elasticity, i.e. the elastic volume variation of a solid under an isotropic (i.e. hydrostatic) stress. It is defined as the ratio of hydrostatic stress to volumetric strain, and is the inverse of **compressibility**.

3.1.2 Plasticity

The elastic domain ends at a so-called **yield stress**, beyond which deformation gets easier, but becomes irreversible, i.e. the original shape is not restored upon stress release. This is the plastic domain. Yield stress is usually defined as the stress for which the stress-strain curve departs from the linear elastic response by a conventional but arbitrary offset strain value, usually chosen to be 0.2%. This is called the 0.2% yield stress. Solids with a large yield stress obviously have an extended elastic domain, such as springs for instance. The yield stress of steel is significantly larger than that of "butter from the window", and its plastic deformation requires temperatures by far higher, but the physics of deformation are fairly similar.

Ductility characterizes the solid ability to undergo significant plastic deformation between yield point and rupture. It can be opposed to ***brittleness***.

Hardness measurements are an alternative characterization of yield stress, used instead of uniaxial tests for practical reasons, but they also incorporate work hardening properties (increase of flow stress with strain) that make result interpretations more difficult. Hardness measurements are often carried out by indenting of the solid under constant load with a very sharp and hard indentor, and subsequently measuring the indentation size.

The stress required to keep the solid deforming at constant strain rate beyond the yield stress is called the ***flow stress***. It usually starts increasing with strain (this is called ***work hardening***). In tensile deformation, it goes through a maximum called ***tensile strength***. Beyond this maximum starts an unstable softening leading to plastic failure.

It is worth mentioning that snow is not a compact solid. As a consequence, the above quantities may be applied (with caution) in the elastic domain, but may sometimes be irrelevant in the plastic one.

3.1.3 Static vs dynamic loadings

If a drum parchment is pressed firmly but slowly, it does not emit any sound, whereas it sounds beautifully if sharply knocked. This is the difference between static and dynamic loadings. The elastic and plastic properties discussed above mainly stand for "quasistatic" loadings, i.e. when the time required to apply the load is longer than the travel time of an acoustic wave throughout the material. In the opposite case of dynamic loadings, the effects on the solid of the propagating wave front, of inertia, and of the associated high stress and strain rates have to be taken into account. This situation is likely to be found in the case of avalanches triggered by pedestrians, skiers, snowmobiles, or artificial triggering by explosives for instance, in which acoustic wave transmission through the snow layer, or at air/snow or snow/snow interfaces, play a key role in the *WL* failure process, whereas quasistatic loadings are usually relevant in natural avalanche release (except in specific cases such as cornice failure for instance). The abruptness of the skier's dynamic impulse is of considerable importance. It is indeed transferred to the weak layer through the propagation of an elastic wave through the slab.

If part of the *WL* has a lower density than that of the layer above (as mentioned in section 2.1), the wave velocity would be smaller in that part of the *WL*. It can be inferred from the conservation of the transferred energy flux that the wave amplitude varies in the opposite way, i.e. is larger in this *WL* zone. This effect may significantly favor its failure, all the more so because this lighter zone is more brittle (section 2.1).

3.2 Fracture initiation and extension

Fracture is obviously influenced by loading geometry. Characteristics of the stress tensor (compression, tension, and shear components) play a key role. But fracture phenomena are (too) often studied in the case of continuous, homogeneous, and defect-free media. Yet, fracture of solid materials involves the nucleation and propagation of a two-dimensional object (a crack) in a three-dimensional one (a solid). This is why geometrical and topological concepts are of interest in this field. A crack has indeed "to know" where to start from! It is indeed difficult to imagine where a crack would first appear in a perfectly homogeneous medium loaded homogeneously.

Inhomogeneous stresses, arising from particular slope shapes or snow layers packing, may act as stress concentrators. This is the case, for instance, at the junction between a lower steep slope and an upper gentler one, at which the top snow cover experiences larger tensile stresses, favoring crack nucleation. This is also the case at interfaces between solids with different elastic or plastic properties, as the interface between a slab and an underlying older snow layer: they are preferential sites for stress concentrations, as under the same stress they would deform differently if they were not stuck together. Taking the problem the other way round, if they cannot deform freely, so-called "incompatibility stresses" develop at the interface, that may help crack nucleation.

A fundamental point is the necessary combination of nucleation and propagation processes. There is indeed a general trend in physics that phenomena requiring some energy are more easily achieved through nucleation of a small nucleus that subsequently and gradually expands to the whole system, rather than from a sudden generalized process that would require too much energy all at once. This is the case for snowflake generation, as mentioned in chapter 2: solidification starts on small dust particles, from which the flakes grow, absorbing more supersaturated water droplets, and gradually aggregating with other flakes in the vicinity. If this was not the case, a huge mass of snow (or ice!) would form in the atmosphere in a very short time, and its fall, though carrying the same amount of water molecules, would probably be slightly more hazardous for people staying below than a longer but gentle snow shower.

In the case of avalanches, release essentially results from the failure of an interface, either between two snow layers for slab avalanches, or between the snow cover and the ground for full-depth avalanches, or through slab thickness for crown crack opening. As above, such a process is more easily achieved starting from a small flaw. In this case, flaw expansion may proceed gradually up to a given size, and then would expand catastrophically. This behavior is found in all fracture problems, and obeys a very simple and quite useful criterion, put forward by the British engineer Alan Arnold Griffith in the beginning of the twentieth century. Its physical origin is fairly easy to understand using a couple of examples, as discussed in section 3.3 hereafter.

3.3 Griffith's criterion

Stretching a rubber band requires energy. This energy is stored in the band in the form of so-called "elastic energy", which can be clearly evidenced when the applied tension is suddenly released (if the band fails and occasionally jumps to your face for instance!).

Let us now stretch a paper sheet between our hands. As for the rubber band, bonds between molecules elongate, but to a lesser extent, due to a higher stiffness. This mechanism also corresponds to some energy storage.

We shall now try to understand how the paper sheet may tear apart. Under constant tension, let us take a knife and cut a small notch perpendicular to the tensile axis. Nothing happens. We slightly increase the notch size. Still nothing. And so on. At a given stage however, for a "critical" notch size (see appendix A), the sheet will suddenly tear apart, releasing the total stored energy. A.A. Griffith published in 1920 his famous criterion (Griffith 1920) that accounts for this phenomenon.

Let us consider a material experiencing a tensile stress σ (taken here as a scalar and not as a tensor, for the sake of simplicity) with a notch of size $2r$. For reasons that will appear obvious hereafter, the "stress concentration (or stress intensity) factor" K is defined as:

$$K = \sigma\sqrt{2\pi r} \qquad (3.1)$$

Griffith's criterion states that the material fails when the stress concentration factor reaches a critical value K_c, i.e. when:

$$\sigma\sqrt{2\pi r} = K_c \tag{3.2}$$

where K_c is the so-called material toughness. For instance, at constant stress σ, $K = K_c$ when the flaw size reaches the critical value $2r^*$ defined by equation (3.2), or at constant flaw size, when the applied stress reaches the critical value σ^* also defined by equation (3.2).

This criterion shows that the larger the applied stress, the smaller the critical flaw size, as intuitively expected.

A thorough demonstration of this criterion can be found in mechanics textbooks, with extensive developments of K_c expressions for different loading types (shear, tension, etc.). However, we are only interested here in the physical basis of the criterion, which can be made clear using very simple "hand waving" arguments, as follows.

Let us consider a paper sheet loaded in tension, that stores elastic energy (in dark grey in Fig. 3.2). The stored elastic energy per unit area of the paper sheet is by definition:

$$\int \sigma \, d\varepsilon = \int \frac{\sigma \, d\sigma}{E} = \frac{\sigma^2}{2E} \tag{3.3}$$

where σ and ε are respectively the stress and elastic strain, and $E = \sigma/\varepsilon$ is the elastic modulus. Obviously, the stored elastic energy is larger for a "soft" material, i.e. with a low E value.

On the other hand, opening a crack perpendicular to the tensile axis needs some "tearing" energy in order to cut bonds between molecules, but in doing so, it relaxes the elastic energy that was stored in the light grey zone in Fig. 3.2, helping the opening process. The light grey zone surface scales as r^2, and so does the elastic energy relaxed in this zone, acting as a driving force for the rupture process.

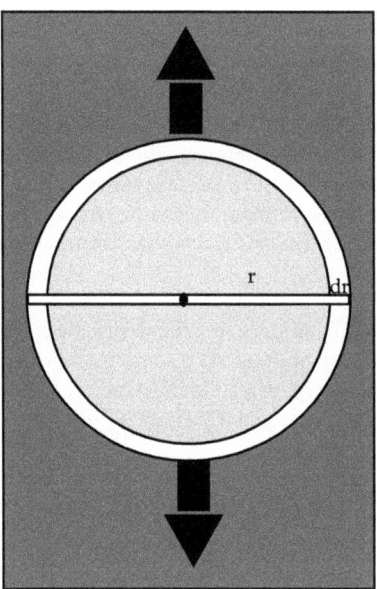

Fig. 3.2 *Sheet of paper with a notch of size 2r. The sheet is loaded in tension, perpendicular to the notch (arrows). Elastic energy is stored in the zone in dark grey, but is relaxed in the light grey one, schematized as a disk of radius r. When the notch size is increased by a small value dr, elastic energy is released in turn in the white zone of area $2\pi r \, dr$, from which the energy balance of eq. 3.5 can be established.*

In comparison, the notch size scales as r, and so does the energy required to open it, that acts as a resistance to notch expansion. The two curves intersect for a critical notch size $2r^*$ (Fig. 3.3), beyond which the driving force exceeds the resistance, leading to dramatic notch expansion and failure.

More precisely, increasing the crack size by a small amount dr on both sides requires an additional tearing energy $2\gamma\,dr$, where γ is the tearing energy per unit length. This energy is partly balanced by relaxation of the stored energy in the white crown (in Fig. 3.2) of surface $2\pi r\,dr$. At a given point, this relaxed energy will be so large that it will overbalance the energy required to extend the notch further. This critical situation is reached when:

$$\frac{\sigma^2}{2E} 2\pi\, r\, dr = 2\gamma\, dr \tag{3.4}$$

or equivalently:

$$\sigma\sqrt{2\pi\, r^*} = \sqrt{4E\gamma} \equiv K_c \tag{3.5}$$

This expression is equivalent to eq. 3.2, where $\sqrt{4E\gamma}$ is identified to K_c. It is worth mentioning that the numerical factor is approximate, due to both the crudeness of the approach and its two-dimensional character.

At this stage, the knife becomes useless, and the notch size will increase spontaneously and dramatically.

This Griffith instability criterion is fundamental in all mechanical fracture processes, and particularly in avalanche release. It plays a key role in weak layer instability and crown crack opening,

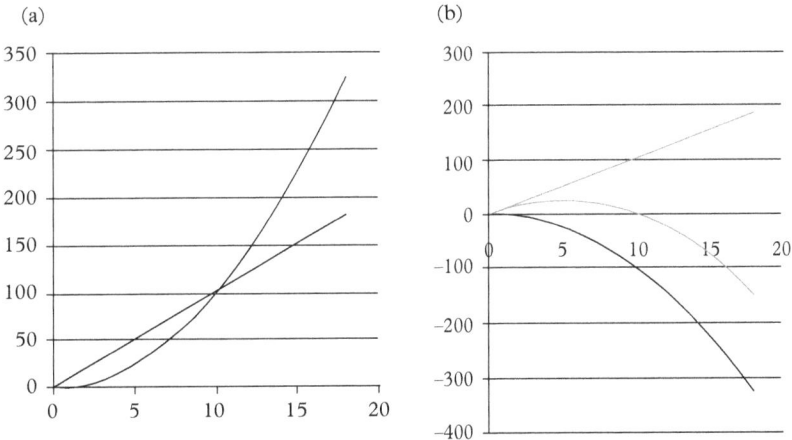

Fig. 3.3 *Origin of the Griffith's critical notch size (schematic curves, arbitrary units). (a) Straight line: tearing energy (scaling as crack size $2r$) vs r. Parabolic curve: stored elastic energy (scaling as r^2). The two curves intersect at the critical point, beyond which the relaxed energy (parabola) exceeds the resistance to crack extension (straight line). (b) The balance between tearing and relaxed stored energies can be represented by the difference between the two curves of Fig. 3.3a, and results in a transition corresponding to the curves' intersection in the top figure.*

as discussed in chapter 5. A typical snow toughness value $K_c = 430$ (+/− 90) Pa m $^{1/2}$ can be found in (Kirchner et al. 2002).

Nevertheless, Griffith's criterion deals with the behavior of an ideal isolated flaw in an infinite solid. In real life, however, and more especially in snow, many flaws may be present simultaneously, and interact. It is intuitive that the system would behave very differently depending on whether we are dealing with an isolated flaw in a continuous solid, or a population of elastically interacting flaws, not to mention a possible evolution towards a continuous array of connecting flaws. This is a critical point, as defined in appendix A. An example will be given in the case of full depth avalanches in chapter 6.

3.4 The brittle to ductile transition

Anybody knows that a block of butter left by the window in summer splashes as a lump of mud if dropped on the ground, whereas it may break into pieces if just taken out of the freezer. Butter is brittle at low temperatures, and ductile at higher ones. This phenomenon is quite general, and is known as the brittle–ductile transition.

The explanation can be given in a simple way on the basis of a combination of Griffith's criterion and of the plastic behavior of the material. Taking the same example as above, the butter from the window is plastic, i.e. it can change shape irreversibly under stress. By contrast, the butter from the freezer is elastic: it deforms under stress (although to a lesser extent than a rubber band), but returns to its previous shape upon stress release. Under larger stresses, it may break in the latter case (it is brittle), or fail plastically in the former one (it is ductile).

Such different behaviors can be illustrated using stress concentration vs time diagrams. Let us consider a brittle material containing a flaw of size $2r$ and gradually loaded at constant stress rate. As long as the notch size does not change, the stress concentration (or intensity) factor K also increases linearly with time, as schematized by the straight dashed line with a slope $dK/dt = \dot{K}$ (Fig. 3.4). The material fails at a time t_{RE} when the straight line of slope $\dot{K}$ intersects the horizontal line $K = K_c$.

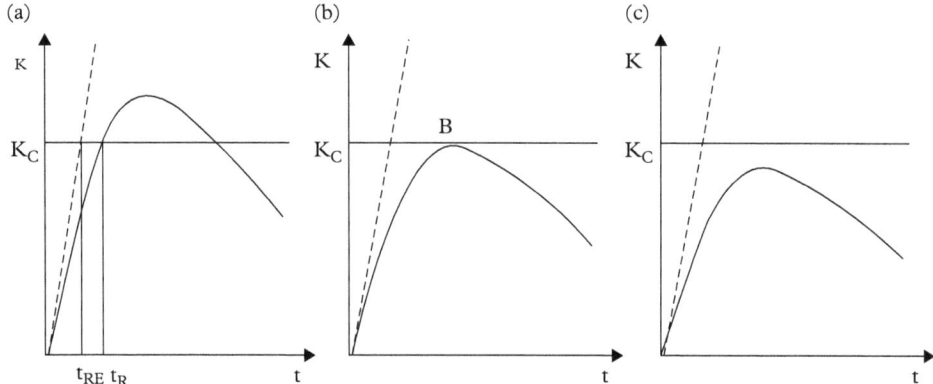

Fig. 3.4 *Stress intensity factor K vs time diagrams. (a) Pure elastic case (dashed line) and brittle plastic case (solid curve) for which the failure time is delayed ($t_R > t_{RE}$). (b) Bifurcation: Brittle to ductile transition. (c) Ductile case: K never reaches the critical K_c value: the material tears apart instead of breaking off.*

We now consider a plastic material. As discussed above, the plastic character means that if stress exceeds a so-called "yield stress", the material deforms plastically, i.e. in an irreversible way. Let us now focus at what happens in the vicinity of the flaw in a stressed plastic material. Crack tips are examples of "stress concentrators", which means that the local stress at the crack tip itself tends to infinity if the tip is acute. It remains large in the vicinity of the tip, and gradually decreases at larger distances. The zone within which the stress exceeds the yield stress is called the plastic zone. It is obvious that the plastic zone size increases for decreasing yield stresses (Fig. 3.5), or in other words as temperature is increased for a given material. Plastic activity in the plastic zone results in a blunting of the crack tip (Fig. 3.6), which in turn reduces the stress concentration.

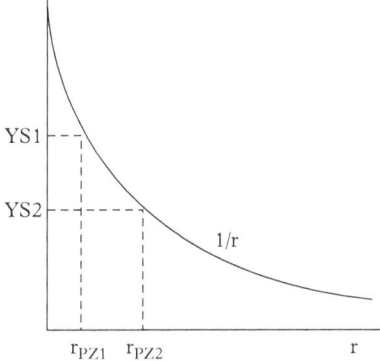

Fig. 3.5 *The plastic zone size r_{pz} is defined as the intersection of the crack stress field, scaling as $1/r$, where r is the distance from the crack tip, and the yield stress level YS. A large yield stress YS1 corresponds to a small plastic zone size r_{PZ1} (short dashes), and conversely for a small yield stress YS2.*

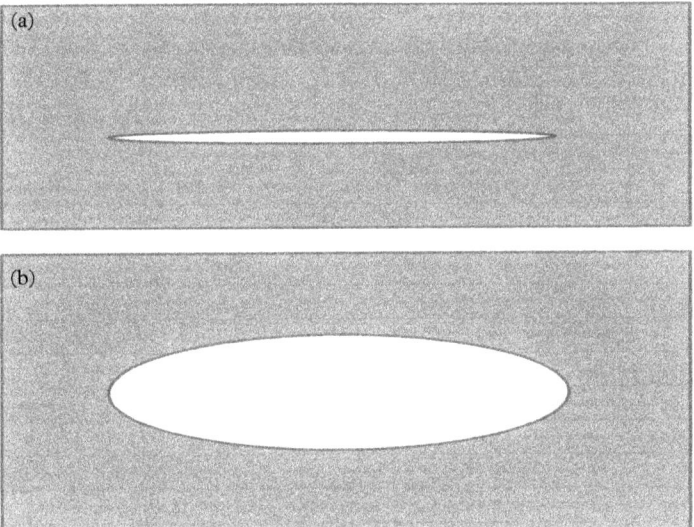

Fig. 3.6 *(a) acute crack, (b) blunted crack.*

As a consequence, instead of following the dashed straight line of Fig. 3.4, the stress intensity factor, in the case of a moderately plastic material, increases more slowly, and the critical value K_c is reached after a longer time $t_R > t_{RE}$ (Fig. 3.4 a). This is usually interpreted by the testing machine (that has a limited scientific education) as a larger toughness. If the material becomes more and more plastic (temperature increase, or larger water concentration in wet snow), the $K(t)$ curve goes through a bifurcation corresponding to the brittle–ductile transition (Fig. 3.4 b), and becomes eventually ductile (Fig. 3.4 c), since it cannot reach the K_c value any more. Beyond the bifurcation, the crack tip blunting is dominant, and the solid tears apart instead of breaking in a brittle way.

3.5 Coulomb's law of friction

Coulomb's law of friction states that the friction between two solids is proportional to the normal load experienced by the contact surface. This point can be easily illustrated by a simple experiment (Fig. 3.7): an empty glass is laid on an inclined board, with an angle α with horizontal. The angle α is gradually increased, and kept just below the angle $α_o$, for which the empty glass starts gliding down (Fig. 3.7 a, b). Now the glass is carefully filled with liquid. Despite the additional weight, the balance is left unchanged, and the glass does not start sliding down (Fig. 3.7 c).

Coulomb's law can be easily explained in the following way: real solid surfaces are not perfectly planar, and when the normal load increases, small bumps at surfaces deform, increasing the actual contact surface between the solids (Fig. 3.8). Under moderate loads, the deformation remains in the elastic domain, i.e. is proportional to the load, and so do both the contact surface, and the related friction force.

For larger loads, however, the elastic response may deviate from linearity, invalidating Coulomb's law. This is why car tyres require high pressures in order to remain in the linear elastic domain, and why heavy cars need wider tyres, for the same reason. This is also why such a friction law has to be applied with caution in the case of snow, which is not a solid in the sense of a dense and continuous material that would exhibit an extended linear elastic domain before deforming plastically. Snow is indeed an aggregate of brittle ice crystals and contains large amounts of air. Under a compressive load, instead of deforming elastically, snow, and more particularly weak layers, may readily collapse in an irreversible way due to ice crystals failure. Snow contact surface with older snow layers or the bedrock turns in this case into a loose medium made of ice debris, opposing a dramatically reduced friction to down-slope slide, in strong contrast with Coulomb's law predictions.

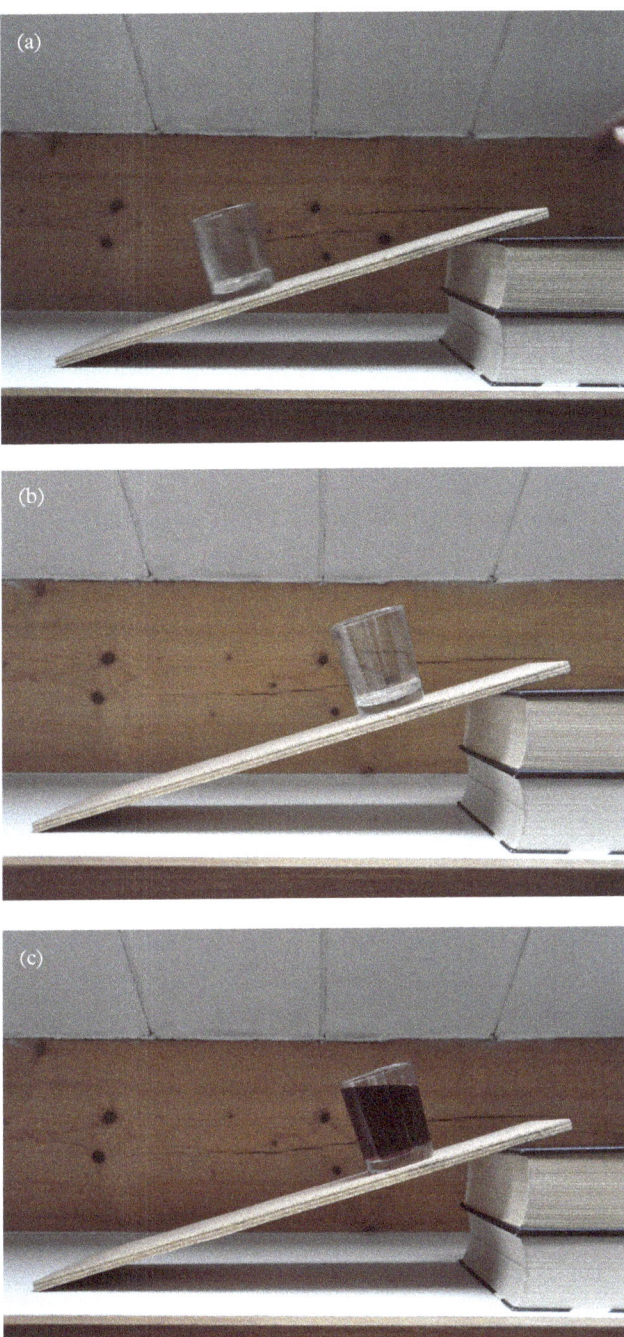

Fig. 3.7 *Home experiment showing the proportionality between the frictional force and the glide driving force in a conventional compact solid. An empty glass is laid on a board inclined just below the exact angle for which the slightest touch on the board would cause the glass to slide down (top: sliding glass, middle: immobile glass). The glass is then filled very carefully. In spite of the additional weight, it remains immobile (bottom): the extra glide driving force is exactly balanced by the increase in friction force. This is usually not the case in snow (see text).*

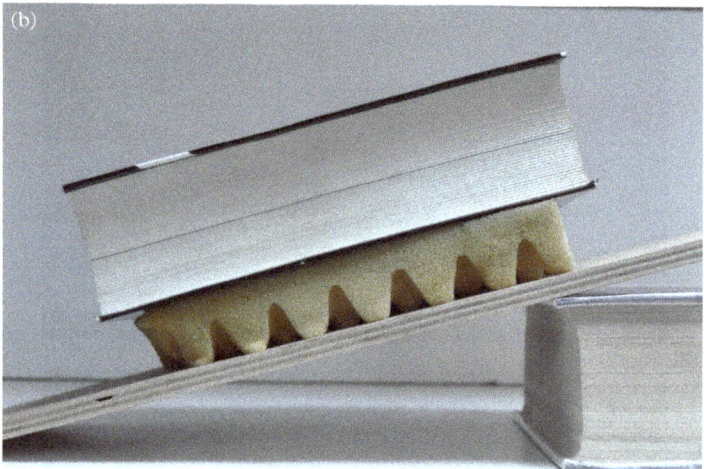

Fig. 3.8 *Illustration of friction between two solids, an elastic material laid on a board. Top: undeformed contact surfaces. Bottom: elastically deformed contact surfaces. In a conventional compact material, the contact area is proportional to the normal stress, and so is the friction.*

4
Slab Avalanche Release: Data and Field Experiments

Who picks up a flower disturbs a star
Théodore Monod

Starting zone sizes are shown to obey statistical laws, named "power laws": the recurrence time of an event of a given size increases in a precise proportion with its size. Extrapolation of such laws fitted on small sized events allows a determination of recurrence times for big and uncommon events.

The key role of the weak layer (WL) failure is illustrated by "Propagation Saw Tests" (PST), showing that the collapse of a WL zone of a few decimeters may act as a switch triggering a very large scale spontaneous WL failure. However, the consequences of such a collapse may be damped down by sintering of broken WL grains.

We analyze bridging indexes, often used to estimate WL resistance to collapse under loading. We define a new bridging index, extending the usual one to the case of elastic bending, and we discuss the validity domains of both of them.

4.1 Geometry and dynamical characteristics

Figure 4.1 shows two examples of slopes after a typical slab avalanche release. The avalanche starting zone is bounded by a crown step at the top, flank steps on both sides (Fig. 4.1 a), remnants of the destabilized slab, and a "stauchwall" (slab bottom boundary).

The crown step formation is the most visible, and sometimes the audible sign of slab avalanche destabilization, reinforcing the old belief that it is the initial event resulting in destabilization of the slab. This is proved to be wrong, in particular by impressive "bangs" sometimes heard during crown crack opening, resulting from sudden relaxation of tensile stresses withstanding the weight of the hanging slab.

The stauchwall may consist of outcrops bounding the slab bottom, or of the bottom part of the slab that has not yet been destabilized during the first steps of slab release. It may interact with the destabilized slab in different ways:

i) Immediately after crown crack opening, the compressive stress on the stauchwall increases; if the stauchwall is strong enough a so-called "toad skin" appears at the snow slab surface, a signature of local buckling.
ii) The flowing slab may also jump over the stauchwall after shearing at the interface, and continue its way, gradually slowing down and eventually stopping on gentler slopes (Fig. 4.1 b, Fig. 4.3).
iii) If made of snow, the stauchwall may also be swept out, preferentially in the case of heavy snows, increasing the amount of flowing material.

Fig. 4.1 *Typical slab avalanche release. (a) Haute Tarentaise, Tignes district, Bec Rouge -Chardonnet, 18 April 2018 17:45. (Photograph by Alain Duclos, data-avalanche) (b) Haute Maurienne, Pointe du Clôt, 11 February 2017. (Photograph by Clément Bléteau, data-avalanche).*

On a non-planar slope, the gliding slab often breaks into pieces. The resulting granular medium flows down, and may transform into a so-called airborne powder avalanche in the case of cold and dry snow.

In many cases no particular remnant signs of an initial failure are visible on the glide surface itself after the slab has flowed. Such a simple geometry may suggest a very simple triggering mechanism. This is probably the reason why in early models it was assumed that the initial failure was the opening of a crown crack, which then expanded sideways into flank ruptures, causing the final slab destabilization. A local perturbation by a skier was often supposed to be responsible for crown flaw opening. Although such a mechanism may be valid in some cases, it is not in the large

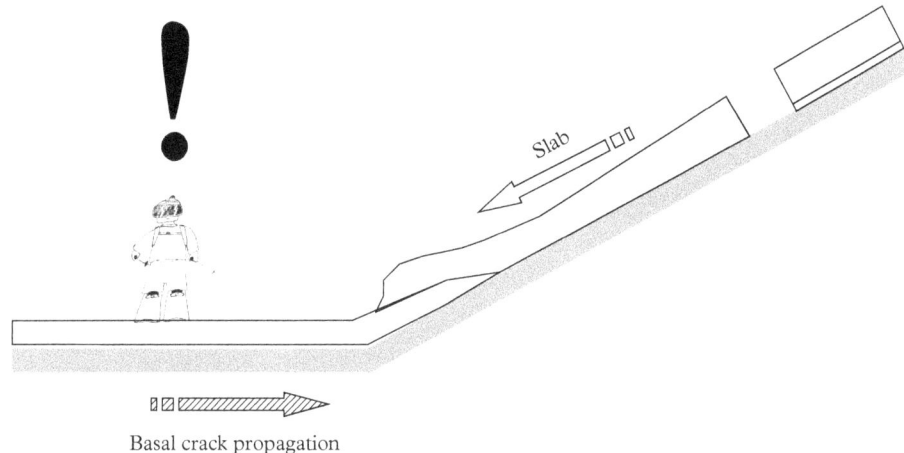

Fig. 4.2 *Stauchwall typical shearing and overcoming by a sliding slab (after Alain Duclos).*

majority of release events. In the general case, it is difficult to decide where the initial failure was located. Understanding this point first calls for a statistical investigation of slab avalanche characteristics, developed in section 4.2.

It appears from Figs 4.4 and 4.5 in section 4.2 that starting zone widths range from a few meters up to 1 km, and thicknesses from a few centimeters up to a few meters. Assuming more or less circular starting zones, the starting snow mass may thus be comprised of anything between less than $1m^3$ and several millions of m^3. Taking a snow density of 300 kg/m^3, the largest starting zones correspond to masses up to one million tonnes. Such an estimate does not take into account the additional snow mass collected during avalanche flow, that may significantly increase the total moving mass up to several millions of tonnes. Slab avalanche velocities may reach 100 to 150 km/h (i.e. about 30 m/s), resulting in kinetic energies of the order of several million kJ for the largest slab avalanches with a snow density of 300 kg/m^3. This impressive figure is comparable to the energy of more than 10,000 cars of 1 tonne crashing into an obstacle at 80 km/h.

Airborne powder avalanches, that are often associated with slab flows, have comparable masses, and may reach velocities as large as 300 km/h (80 m/s), associated with an increase of kinetic energies up to one order of magnitude.

4.2 Statistical aspects: scale invariance

As mentioned above, at first glance, snow slab release processes may be thought to result from simple deterministic mechanical rules. In this respect, a precise knowledge of the geometry and size of starting zones may be of high interest. In addition, they may be used as input parameters in avalanche flow simulations. However, field measurements show considerable variability, suggesting a preliminary statistical approach.

For this purpose, a database of more than 5000 avalanches was built up and analyzed (Faillettaz et al. 2002, Faillettaz 2003). Data were recorded in La Plagne and Tignes ski resorts during 3 winters. They contain lots of valuable information such as avalanche triggering modes (artificial,

28 *Snow Avalanches*

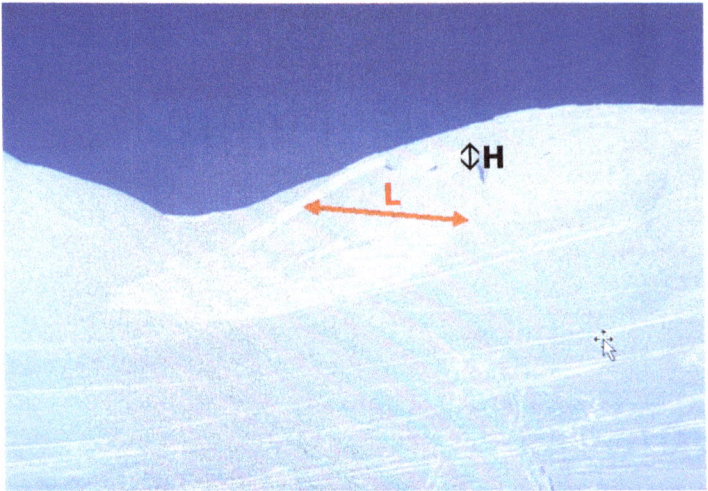

Fig. 4.3 *Snow slab avalanche. The starting zone size is characterized by its width L, and the slab thickness H at the crown rupture. (Photograph by Michel Caplain).*

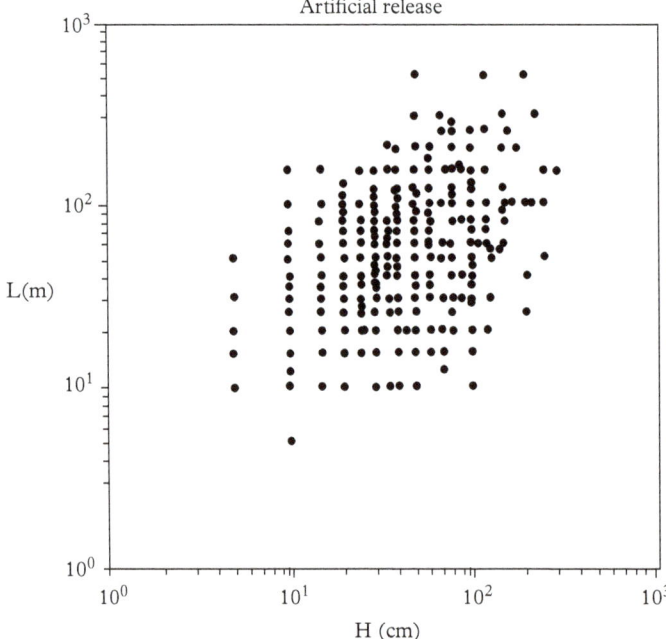

Fig. 4.4 *L and H distributions for 3450 avalanches recorded at La Plagne and Tignes ski resorts (France). Each single point may correspond to one or several events. (Faillettaz 2003, Faillettaz et al. 2006).*

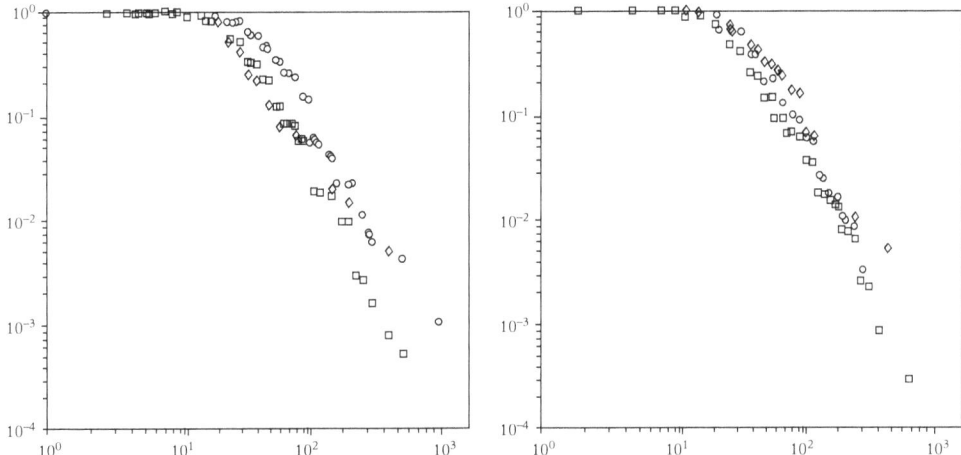

Fig. 4.5 Cumulative distributions of L and H. (a) $N(L>L_0)/N_{tot}$ vs L(m), (b) $N(H>H_0)/N_{tot}$ vs H(cm). Data are retrieved from both artificial and natural triggering recorded at La Plagne and Tignes ski resorts (France) (logarithmic scales). **Squares**: artificial releases, La Plagne; **circles**: artificial releases, Tignes; **diamonds**: natural releases, La Plagne. Both distributions give a similar exponent (−2.4), corresponding to an exponent (−2.2) if expressed in terms of probability density (i.e. non-cumulative distributions) of starting zones areas (see Figs A.3 and 5.4). The saturation at low sizes arises from the fact that L values less than 10m and H values less than 10cm are not recorded. (Faillettaz 2003, Faillettaz et al. 2006).

accidental, or natural), location, etc. In addition, crown crack heights H and lengths L (Fig. 4.3), were recorded for every avalanche, defining starting zone sizes.

In contrast with a current belief, a careful analysis (Faillettaz 2003, Faillettaz et al. 2006) showed that H and L values do not exhibit any correlation (Fig. 4.4).
But behind such an apparent randomness, statistical distributions of each of these parameters exhibit a remarkable feature. In a log-log plot, they align on a straight line with a negative slope (−*b*) (Fig. 4.5). In other words, they obey power-law distributions, also known as "scale invariant" distributions (see appendix A). The exponents are around 1.4 ± 0.1 for non-cumulative distributions, i.e. 2.4 ± 0.1 for cumulative ones (see appendix A, Fig. A.3). All available avalanche data, from both natural and artificial triggering, align on the same power law, whatever the winter season, the mountain range, the slope orientation, or the gully they start from.

Such statistical laws are similar to those describing earthquakes (Gutenberg and Richter 1956), but also landslides, rock-falls, etc. (see appendix A). This remarkable law allows an easy prediction of large avalanche occurrences, through an extrapolation to large scales of the scale-invariant distribution fitted on more frequent small-scale events.

4.3 The weak layer, starting point for slab avalanche release

Despite the fact that slab avalanches may occur spontaneously, they are often triggered by skiers or by explosives or other blasting devices, which is not the case for full-depth avalanches as will be discussed in chapter 6. Such a characteristic strongly suggests that *WL* failure is a key step in slab avalanche release.

As mentioned in chapter 2, a *WL* consists of granular aggregates of polyhedral ice grains bonded by brittle ice bridges. They can easily collapse as a house of cards, even under a moderate but brief loading, such as a skier impulse, and transform into a fluid-like layer. A similar collapse takes place during skiing on surface hoar, the resulting loss of cohesion being responsible for the incomparable ski glide feeling experienced in such conditions. It can be imagined that a slab may easily slide down on such a collapsed layer.

Such observations are quantitatively backed up by several tests, as follows.

4.3.1 Propagation Saw Tests (PST)

WL failure actually results from a combination of vertical subsidence (by collapse) and slope-parallel shear (see hereafter). In "usual" conditions, the critical Griffith's crack size is significantly smaller for subsidence than for shear, of the order of a few decimeters in the former case, and of a few hundred meters in the latter, as shown by theoretical calculations (Heierli et al. 2008a, 2008b, 2010). This is intuitively obvious for "moderate" slopes: the work required for collapsing a given area of *WL* is indeed more readily provided by a vertical displacement than by a slope-parallel one. The expansion of the collapsed zone (also named a basal crack) becomes therefore spontaneous (i.e. unstable) for a size of a few decimeters.

Driven by slab weight, such cracks can extend to large distances beneath slabs, and lead in some conditions to avalanche triggering. Since they can also be initiated on horizontal terrain, avalanches may be released during skiing or walking on such zones, provided they are close to significantly slanting slopes. Horizontal terrain is not as safe as usually believed.

Several types of tests were designed in order to further understand how the *WL* collapse may trigger a slab avalanche, as the Rustchblock test (Jamieson and Johnston 1993), high speed photography on skier-tested slopes (Van Herwijnen and Jamieson 2005), and the Propagation Saw Test (PST) (Gauthier and Jamieson 2008).

We shall focus here on the PST, which seems to provide the most precise and reliable observations and quantitative data. This test was designed independently in Canada (Gauthier and Jamieson 2008) and in Switzerland (Sigrist and Schweizer 2007).

Using a snow saw, a parallelepiped block of snow is isolated from the slab, one of the sides being left accessible to observations (Fig. 4.6). The *WL* is gradually cut, starting for instance from the bottom in the case of (Van Herwijnen and Heierli 2009). For a critical cut size, *WL* failure occurs and propagates (uphill in this case), in qualitative agreement with Griffith's criterion (chapter 3). These tests were carried out with slope angles between 28° and 35°, and black markers were placed on the observable vertical snow wall, in order to follow the slab motion during *WL* failure. Despite the fact that this work was aimed at estimating a crack-face friction coefficient, assuming a Coulomb-type friction, several interesting features are observed:

i) The measured critical Griffith's size (i.e. the cut size) for crack expansion is of the order of a few decimeters, in agreement with theoretical predictions mentioned above. This result suggests that the local impulse of a skier or the footprint of a hiker may be responsible for a generalized *WL* collapse failure.

ii) A vertical displacement takes place in the very beginning of the collapse, for the same reason as in i).

iii) This displacement rapidly transforms into a slope-parallel shear motion after propagation of the fracture through the entire sample, since the *WL* is entirely collapsed, and the slab and substrate are in close (though not cohesive) contact.

iv) the expansion velocity of the collapsed zone can be estimated, of the order of 20m/s in the conditions of this particular (but typical) set-up.

Fig. 4.6 *Example of a PST experiment. The weak layer located beneath a long isolated slab block is cut using a snow saw, resulting in crack initiation and spontaneous expansion (after Alain Duclos, data-avalanche). See also videos by Alain Duclos: https://youtu.be/qLr1R-kBpIc, https://youtu.be/H_k5hP1fzh4*

v) in some tests, the destabilized slab decelerates and comes to a rest after a displacement of a few centimeters. This particular behavior seems to preferentially occur for the lowest slope angles.

Such results will be discussed on a different basis in chapter 5. A particular emphasis will be given to point (v), which we will show is the result of a sintering (clotting) mechanism when the *WL* shear strain rate becomes less than a given threshold, for instance for low slope angles. In this case, the triggering process fails, transforming into a simple "whumpf".

4.3.2 Sintering (clotting) experiments

In order to characterize more precisely the *WL* behavior and its possible consequences on avalanche release processes, field experiments were carried out between 2009 and 2013 in the Orelle ski resort (French Alps) (Duclos et al. 2009, Louchet et al. 2013). Shortly after being collected from the *WL* with a shovel, the material was observed to behave as a granular slurry, hereafter named fluid (*F*), made of grains flowing like dry rice, confirming that a mechanically disturbed *WL* may possibly act as an easy glide surface for the overlying slab. However, when left undisturbed for a few seconds, the fluid clotted into a solid (*S*), and got stuck to the shovel (Fig. 4.7). When mechanically disturbed again (e.g. by a mechanical shock), the *S* phase turned back to the *F* one if the disturbance was strong enough, and flowed out from the shovel.

Another version of this test (although not really quantitative at this stage) is shown in Fig. 4.8: a bottle whose bottom has been removed is fastened upside down to a thin and flexible rope, and filled with *WL* snow. After a few seconds, the lid is carefully removed. The snow does not flow out of the bottle. The bottle is then hit by a pendulum made of a metallic mass fastened to another

Fig. 4.7 *The shovel test: a) snow is collected with a shovel from the WL, b,c,d) after a few seconds, snow has sintered, and stays stuck on the shovel while it is gradually turned upside down. (Photograph by Alain Duclos: https://youtu.be/isJhAsL9vpA).*

flexible rope. When the mass kinetic energy is large enough, the bottle content transforms into a fluid and flows out.

A more sophisticated set-up of this test may provide an estimate of the cohesion energy of the clotted snow. Such a sintering (or "clotting") process will be shown in chapter 5 to drastically control slab avalanche release.

4.4 Stability and bridging indexes

It should be interesting to characterize the stability of a snow slab lying over a *WL* using a reliable and user-friendly test. Several tests and corresponding indexes were successively proposed.

A simple stability index was initially put forward (Roch 1966, Perla 1978; https://doi.org/10.1016/B978-0-444-41507-3.50030-1) as the "ratio of the strength of a bed layer to the load on it". Basically, failure is expected to take place when the load exceeds the strength. This index was shown to be often unreliable in predicting most avalanche triggering events.

Several refinements were subsequently proposed, some of them introducing slab hardness, defined as the resistance to penetration. The Bridging Index (*BI*) was thus defined as the "uniform hand hardness index" multiplied by the layer thickness (Föhn 1987, Jamieson and Johnston 1998, McClung and Schweizer 1999). Such stability indexes nevertheless suffer several inconsistencies.

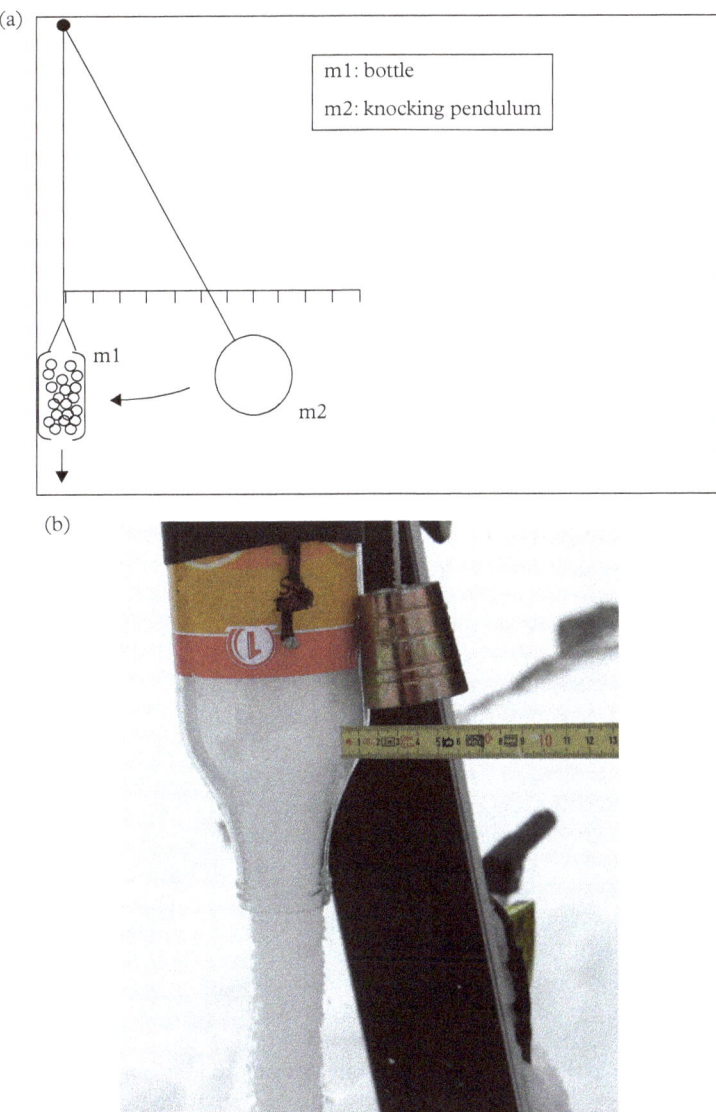

Fig. 4.8 *Solid-fluid transition illustrated by the "bottle test". Top: schematic. Bottom: experimental set-up. (Photograph by François Louchet; https://www.youtube.com/watch?v=bOXYxADasbU).*

One is that they deal with continuous and defect-free media, ignoring the role of stress concentration at possible crack tips (chapter 3 and (Griffith 1920)).

A second one is the definition of the "layer thickness", due to the fact that the actual slab thickness can only be defined after release. In addition, such indexes do not take into account the key role of *WL* collapse (a feature that was actually not established at that time!).

A third (and important) one arises from some kind of mixing up between hardness and stiffness (see chapter 3). It is intuitive that stiff (and not necessarily hard) layers create a bridging effect that spreads out laterally the additional load or impulse, henceforth reducing the "penetration depth" of elastic strain, and the possibility of triggering a *WL* collapse. The finite element model proposed in (Jones et al. 2006), despite the fact that it was focused on shear stress only, was developed on this basis.

Let us now examine the problem from a new view point. In mechanics of solids, the beam deflection theory shows that the deflection of a beam roughly scales as W/Eh, where W is the load, E the elastic modulus (stiffness), and h the beam thickness. For a given skier (or snowmobile) local overload, a reasonable "Flexural Bridging Index" (tentatively named *FBI*) should therefore be the product (Eh) of the elastic modulus E by the layer thickness h. Stiff and thick slabs are hardly flexible, and less likely to damage the underlying *WL*. Slabs should be considered as insensitive to accidental loadings for index values larger than a critical threshold, to be determined by experiments for each type of loading (skier, snowmobile, etc.).

The use of the Bridging Index (*BI*) mentioned above, and defined as the "uniform hand hardness index" multiplied by the layer thickness ($BI = Hh$), should thus be restricted to estimate how the slab resists a local columnar punching that may yield *WL* collapse. Strictly speaking, this has nothing flexural. By contrast, the Flexural Bridging Index (*FBI*) deals with long-range beam elastic bending. Both indexes have indeed different physical meanings, and characterize different situations and different physical phenomena as detailed in chapter 3. *BI* is related to hardness, i.e. plastic (or in some cases brittle) properties, whereas *FBI* uses elastic ones. There is no obvious correlation between them. A stiff material (large E) resists elastic reversible deformation, whereas a hard one resists irreversible penetration (usually collapse of snow crystals). One can imagine that a snow exhibiting quite a stiff flexural response may fail in a hardness test, and conversely.

The operation of *BI* tests is user-friendly. This is less obvious for *FBI* ones, as measurements of the elastic modulus E require acoustic equipment not currently available in skier backpacks. But there is a good correlation between elastic modulus and snow density (Shapiro et al. 1997), at least in dry and cohesive snows, that suggests a new definition of the *FBI* as the product (ρh) of density by slab thickness, to be experimentally calibrated. It is worth noting that correlation between hardness and density is far less obvious (Shapiro et al. 1997), which confirms that stiffness and hardness have quite different physical meanings.

Both *BI* and *FBI* should therefore be used with caution, depending on snow properties. In practice, in order to test *WL* stability, both *BI* and *FBI* should be combined, for the following reasons. Figure 4.9 schematically shows a stiffness-hardness map. For a given slab thickness, using *BI* alone would suggest a good stability at large values of hardness H, i.e. on the right side of the map (at points A and A'). Yet, the actual stabilities at A and A' are different: starting from A, a local decrease of E at constant H ($A \rightarrow B$ trajectory) would not affect stability, but a similar E decrease starting from A' would bring the system into the unsafe zone, due to a lower stiffness. The same argument would apply using *FBI* at constant E values, but different H ones.

In summary, the top-right zone of the map should therefore be considered as safe, and the bottom-left one as unsafe, but if one of the two values of E or H is small, *BI* and *FBI* should be combined to improve *WL* stability testing.

Two other tests may also be used to estimate snow pack stability: the "Tap Test" (https://www.youtube.com/watch?v=6qFOuBYTvbk), or the more sophisticated "Extended Column Test" (*ECT*). They should not be considered as providing bridging indexes, since they are performed on a snow column isolated from the surrounding snow cover: such boundary conditions totally relax long-range elastic interactions that are essential in the elastic flexural response. For instance, the *ECT* requires digging a snow pit down to the supposed *WL* location, and then isolating a

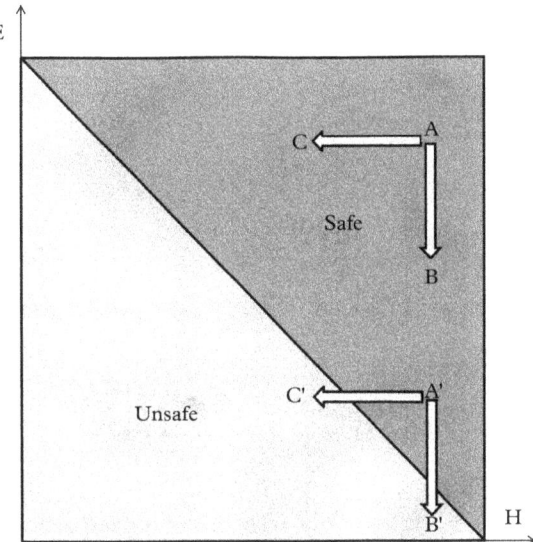

Fig. 4.9 *Schematic stiffness E vs hardness H map. At constant slab thickness, hard <u>and</u> stiff slabs are likely to protect WL integrity, whereas soft <u>and</u> flexible ones may not, defining a safe region (top right) and an unsafe one (bottom left). BI used in the safe (top right) zone but close to the boundary should be validated by FBI, and conversely.*

column from the remainder of the snow pack using a saw or a rope. A shovel is put on one side of the block, and loaded by successive taps until *WL* collapse initiates, and possibly propagates through the entire block width. It provides information on *WL* propensity to collapse initiation and extension, and may be considered as complementary to the *PST*.

Much simpler is the Ski Pole Penetration Test (*SPPT*) (https://youtu.be/1EC9yivm2Kk). It is probably the easiest way to detect the <u>presence</u> of weak layers at moderate depths. It consists in feeling (or measuring) discontinuities in the penetration force necessary to push a ski pole vertically into the snow pack, either in the normal orientation, or upside down in the case of hard snows. However, such tests do not provide any reliable information about the *WL* propensity to collapse.

5
Slab Avalanche Modeling

> *Physics is looking for simple invisible behind complicated visible*
> Jean Perrin

The chapter starts with a review of a few unfounded arguments sometimes used to account for snow slab instability, and often resulting from the application of mechanical laws that are invalid in a granular brittle and healable material like snow.

Statistical aspects are investigated using a two-threshold cellular automaton, one for basal instability, and the second one for crown crack opening. The results reproduce the power-law size distribution of starting zone sizes mentioned in chapter 4, and validate a "4-step" triggering scheme made of successive initiation and expansion events for both the basal crack and the crown crack.

The possible sintering of collapsed WL is then analyzed. It is shown to flow as a slurry for shear strain rates larger than a predetermined threshold, or to sinter in the opposite case, which provides a "joker" to any successful "4-steps" scheme, turning an incipient avalanche into a simple "whumpf".

5.1 Old myths and beliefs to shoot down

As suggested by Figs 4.1 and 4.3, slab avalanche release is widely acknowledged to result from glide of the slab on a snow substrate made of older snow. However, the initiation of such a destabilization is sometimes ascribed to one of several mechanisms that may be qualified as untenable beliefs or myths, as detailed hereafter.

i) A classical schematic in a number of works shows a perfectly cohesive and planar slab/substrate interface. In real life, however, as shown in chapter 3, failure of such an interface as a whole is by far more difficult than if it is initiated from a preexisting flaw, or a flaw created by an external and local loading from a skier or an explosive device for instance. If the flaw size exceeds Griffith's critical value (see chapter 3), it may subsequently expand, driven by the slab weight.

ii) Still in the previous scheme of a defect-free interface, it is sometimes argued that slab avalanche triggering may result from an increase of the additional slope-parallel weight component of a skier, which would break the balance between the driving force for slab glide and the glide resistance (Fig. 5.1).

However, hanging an additional coat on a coat rack would not necessarily disturb the existing equilibrium and break the hook, as long as the resulting load remains below its failure limit. A skier of about 80kg might hardly trigger an avalanche 10,000 times heavier (several thousand tons). If such a statement was true, it would mean that triggered avalanches would have been already very close to the breaking point before experiencing the skier's perturbation, and would have otherwise started spontaneously, probably a few minutes later.

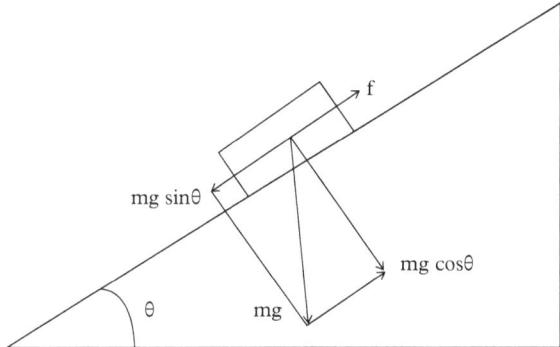

Fig. 5.1 *A "classical" representation of slab mechanical equilibrium on a slope. The slope-parallel component of the slab weight mg is mg sinθ, and the slope-normal component is mg cosθ. The driving force mg sinθ is balanced by the resistance force f. Yet, if Coulomb's friction law was valid in this case, an increase of the slab weight mg would not disturb such an equilibrium (see text).*

iii) It is also implicitly assumed in the literature that Coulomb's friction law (section 3.5) is valid for snow. This would be an additional reason for invalidating the above argument. The additional load and increased friction determined by Coulomb's friction law would indeed exactly balance each other, since the additional weight raises in the same proportions the slope-parallel component (glide driving force) and the slope-perpendicular component proportional to Coulomb's friction. But it was shown in section 3.5 that Coulomb's friction law is anyway hardly applicable to snow, which is a non-dense and collapsible material. The skier's action is obviously an unavoidable parameter in the artificial triggering problem, but it must be introduced in quite a different way than a simple weight addition. As suggested in section 4.3.1., triggering results from a local collapse of the weak layer due to the skier's impulse, that propagates along the slab/old snow interface, drastically reducing the glide resistance of the slab. The mechanical equilibrium of Fig. 5.1 is usually not broken by an increase of load, but by a decrease of friction.

iv) Another frequent statement is that slab avalanche risk is negligible for slope angles less than 30°. Such a statement is at least ambiguous, as two slope angles have to be considered, one at the skier's position, and the other one at the place where the incipient triggering phenomenon becomes unstable, paving the way to final avalanche release. If statements i) and ii) were true, no explanation could be found for "remote triggering", and more particularly for avalanches remotely triggered by skiers or trekkers on a flat terrain, which are actually observed. We shall see in appendix B that *WL* collapse may be initiated on horizontal terrain, but may propagate up to steeper slopes where conditions for slab release are met.

v) Is walking less dangerous than skiing? This is usually wrong. As skis apply the weight on a larger surface than boots, the applied pressure on slabs is lower for skiers than for pedestrians. Thus, skis participate in some kind of "bridging effect" (see section 4.4). Such a statement has nevertheless to be tempered by the fact that, due to a higher velocity, a skier's impulse may have a larger dynamical character, which may in some cases overbalance a lower static pressure.

5.2 Basis for modeling

As mentioned above, slab avalanche release has been studied for a long time using the mechanics of defect-free continuous media. More recently the role of defects was introduced through basic fracture mechanics approaches (McClung 1981, Louchet 2000, 2001) applied to a basal flaw (i.e. located between the slab and the underlying snow substrate), named the deficit zone or basal crack.

In such models, the stability of basal cracks was investigated under shear loading only, essentially based on Griffith's concepts (Griffith 1920, Louchet et al. 2002) or on more complicated but equivalent ones (McClung 1979, 1981, Bazant et al. 2003).

However, again, the weak layer is a non-compact medium that may easily collapse under loads having a compressive component, as already confirmed some time ago (Johnson et al. 2000, Jamieson and Schweizer 2000) and clearly shown recently using Propagation Saw Tests (*PST*) (Gauthier and Jamieson 2006, 2008, Heierli et al. 2008a, 2008b) (see chapter 4).

Basal cracks are therefore initiated by a local combination of weak layer collapse and shear failure modes, varying between the two limiting cases of pure shear and pure collapse (Heierli et al. 2008a, 2008b).

It can be inferred from field observations (Duclos et al. 2009) that such a combined collapse-shear failure of the *WL* ends up with a "fluid-like" layer prone to down-slope glide. The question is whether this fluid phase will remain fluid and allow avalanching, or readily transform into a solid phase, which would stop the triggering process.

However, the main problem that arises in the study of such natural phenomena is twofold: i) numerous parameters are involved, each of them being hardly accessible with reasonable accuracy, and ii) the "initial" state of the system is not exactly known: snowpacks never stop evolving due to meteorological conditions. Avalanches share this characteristic with other natural catastrophic events (in the real life sense as well as in the mathematical one), as landslides, rock-falls, turbidites, and also earthquakes. The statistical approach of the triggering problem developed hereafter in section 5.3 will show that all these phenomena belong to a so-called "universality class", i.e. obey quite similar statistical laws. Another outcome of such approaches is that they provide quite interesting results in spite of the very small number of initial assumptions. As a consequence they may help in designing more elaborate deterministic models, as developed in section 5.4.

5.3 Statistical approach: Playing with cellular automata

Cellular automata were introduced by the mathematician John Neumann in the 1940s. A cellular automaton is a grid of cells, usually (but not always) two-dimensional, where each cell has a state belonging to a set of states. Time is discrete, and the state of each cell at every instant is a function of its state and the states of its neighbors at the previous time step. The most popular example is the "Game of Life" introduced by Conway in the 1970s. (https://en.wikipedia.org/wiki/Conway%27s_Game_of_Life). Cellular automata are now increasingly used as simulation tools.

We designed a specific cellular automaton in order to better understand slab avalanche triggering mechanisms (Faillettaz et al. 2004). The board represents both the weak layer (*WL*) and the slab, seen from above. The load experienced by a cell defines its state, represented by different colors, increasing from dark blue to blue, green, yellow, and red. The red state corresponds to the pre-established cell failure threshold. We start with all cells set in the same state. Snowflakes are then allowed to fall down at random on the board, gradually and randomly changing the cell states from dark blue to blue, green, yellow (Fig. 5.2 and video). At this stage, "load" remains a generic concept, no difference being made between shear or compression components of the stress tensor. The exact nature of *WL* failure is not specified, and may result from either pure shear (as in full-depth avalanches, see chapter 6), or as a combination of subsidence and shear, characteristic of slab avalanches (see *PST* in chapter 4). When a cell turns to red, it fails, and its load is redistributed on its neighbors. If one of them was already yellow, i.e. close to the *WL* threshold, it becomes red and

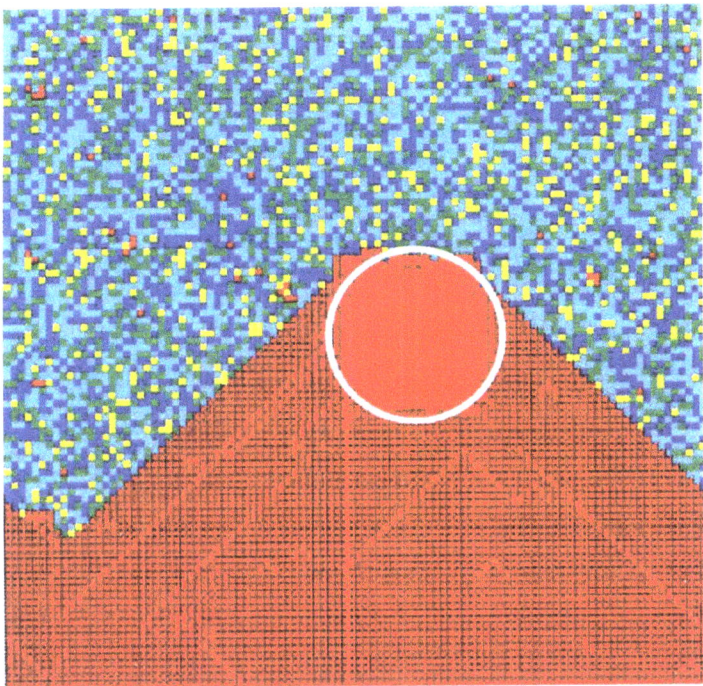

Fig. 5.2 *http://jerome.faillettaz.free.fr/AC_animation.htm*
Spontaneous triggering. The load experienced by the WL is gradually increased by load increments ascribed at random to individual cells. A cell receiving a load increment changes its state by one unit. When one of them reaches the "red" state, it fails and redistributes its load to neighbors. A critical state is eventually reached, for which the failed zone extends catastrophically. The white circle shows the (plain red) avalanche starting zone. (Faillettaz et al. 2004).

fails in turn, and so on. This is how a collective behavior appears in cellular automata. Such rules yield a behavior quite similar to chain reactions that take place during domino avalanches or nuclear fission, and eventually lead to a critical state for which the collapse would extend to the whole board.

But this phenomenon alone does not fully describe avalanche release. This is the reason why a second criterion is introduced to account for crown crack opening. The algorithm analyzes tensile stresses between neighbor cells, i.e. load differences between cells of different colors. When such tensile stresses reach a second pre-established threshold, the corresponding cells separate from one another, initiating a crown crack in the slab, that becomes unstable and opens after several computer steps, triggering the avalanche (Fig. 5.2 and videos).

These simulations clearly illustrate Griffith's rupture criterion discussed in chapter 2. They can also simulate artificial triggering (Fig. 5.3 and video).

The starting zone sizes can be easily identified (Fig. 5.2). Their statistical distribution is obtained analyzing a large number of simulation runs (Fig. 5.4). It exhibits a "power law distribution" (see appendix A), characterized by a straight line with a negative slope on logarithmic plots, in perfect qualitative agreement with field measurements (chapter 4). The slope of this line is the power law exponent. For the sake of simplicity, the exponents dealt with here are expressed in

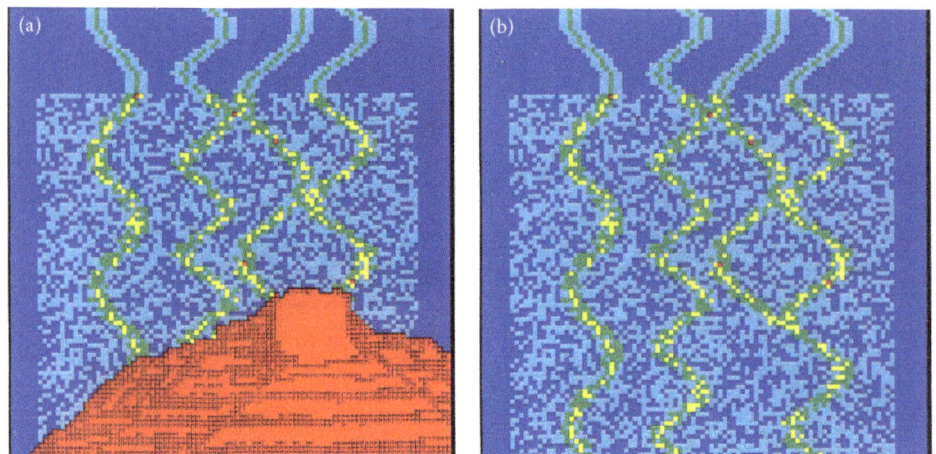

Fig. 5.3 *(after Faillettaz and Louchet).*
http://jerome.faillettaz.free.fr/AC_animation.htm
Skier-triggered avalanche obtained by the cellular automaton. Skiers come from the top of the figure. The direct damage produced by skiers is supposed to extend up to 3rd neighbor cells (left figure). The triggering process starts from a place where two skiers' tracks slightly overlap each other (left figure), and spontaneously extends to the plain red zone (right figure), allowing avalanche release (Faillettaz et al. 2006).

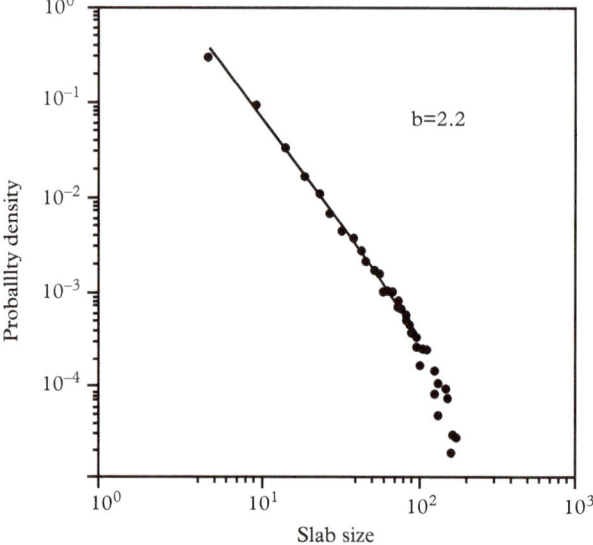

Fig. 5.4 *Typical probability density (i.e. non-cumulative distributions) of slab avalanche starting zone areas obtained from the cellular automaton. Avalanche field data (b = 2.2) (chapter 4) are reproduced if $\alpha = 0.5$ (see also Fig. 5.5 and appendix A, Fig. A.3) (Faillettaz et al. 2004, 2006).*

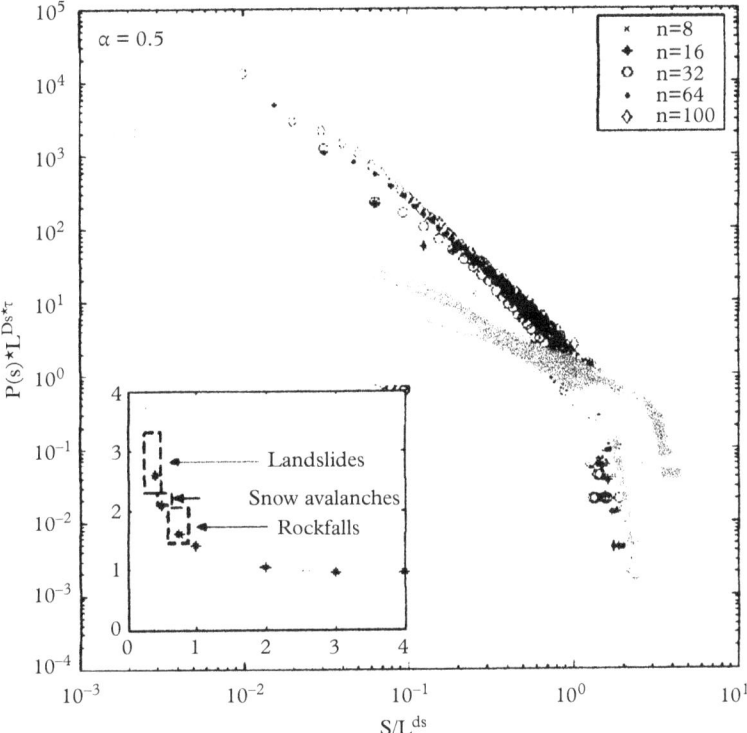

Fig. 5.5 *Probability density of starting zone areas for various types of gravitational flows, given by the cellular automaton as a function of α values. Insert shows the variations of the exponent (curve slope) with α (Faillettaz et al. 2006).*

terms of probability distribution functions of areas, in order to easily compare them with other types of gravitational flows.

Quite similar distributions are indeed observed for landslides, rock-falls, earthquakes, etc. They are also found in various other geophysical phenomena, the Gutenberg–Richter law in the case of earthquakes (Gutenberg and Richter 1956) being a well-known example.

The exponents b given by the automaton can be varied by tuning a single parameter, defined as the maximum value of the ratio of slab rupture to weak layer failure thresholds (Fig. 5.5):

$$\alpha = max\left[\sigma_0 / \tau_0\right]$$

In order to fit simulation outputs to observations, a is always found smaller than 1 (Fig. 5.5). The b exponent measured values are 1.75 ± 0.3 for rock-falls, 2.2 ± 0.1 for snow avalanches, 2.8 ± 0.5 for landslides, and are reproduced for a values of 0.6–0.9, 0.45–0.55, and 0.2–0.5 respectively. a values closer to 1 (i.e. smaller b values) correspond to a more isotropic cohesion. This seems to be the case for rock-falls, as compared to avalanches, and even more to landslides, that involve more layered materials. The smaller b values for rock-falls (i.e. lower slopes in log-log distributions) favor a relatively higher number of large size events, in contrast with landslides.

Such results are reminiscent of the seminal "sand pile model" by Bak et al. (1988), at the origin of the concept of Self-Organized Criticality (SOC), defined in appendix A. A specific condition is required for SOC phenomena: the loading rate must be low enough to avoid avalanche overlapping. In other words, a given avalanche must not be released as long as the previous one is still active. As the automaton is run independently for each avalanche, successive avalanches do not overlap, strongly suggesting that the SOC requirement is fulfilled. This is also the case in avalanche starting zones data mentioned in section 4.2., to which the automaton results are compared, because they are recorded in different winters, and different gullies or mountain ranges.

The situation here is nevertheless slightly different from Bak's model in various aspects. In particular, we are dealing here with simulations of snow slab avalanche release, that require the introduction of a second threshold ascribed to crown crack failure, whereas Bak's simpler model may be more suitable for loose snow avalanches (chapter 6).

Such cellular automata validate a triggering scheme made of successive initiation and expansion events for both the basal crack and the crown crack. However, they do not incorporate any healing. If such a mechanism may be ignored in fast enough processes, sintering (clotting) experiments mentioned in section 4.3.2 suggest that they may play a key role in the two first steps of basal crack nucleation and expansion. This point is dealt with hereafter.

5.4 Sliding or sticking?

The above analysis shows a clear and logical succession of events inevitably leading to avalanche release. However, do you remember this strange feeling that "it" would have released, but it did not? Or when the first skier started crossing the slab very carefully, the second as well, that everything seemed to be OK, and suddenly the whole slab failed? The answer to this "mystery" is based on the fact that, as the triggering process results from various successive steps in series, if a single one is missing for some reason, the whole chain is broken, and nothing happens, at least for a while.

As described above, the weak layer (WL) may collapse, for instance due to a skier's impulse, into an unstable material prone to glide. The size of this initial collapse is usually larger than the critical collapse instability size. This size has been indeed estimated around a few decimeters either by theoretical calculations (Heierli et al. 2008a), or using *PST* tests. It may be slightly larger in real life, since both *PST* tests and theoretical calculations deal with an isolated snow block, with free side surfaces.

As a consequence, the collapsed zone would be expected to immediately extend throughout the WL, leading to avalanche triggering. Yet, one may wonder why lots of collapses only result in simple "whumpfs", without any release!

The explanation is as follows. As mentioned in chapter 2, skiing on surface hoar provides incomparable glide feelings. On the other hand, it is well known that stopping for a few seconds in some "warm" snows results in snow sticking on the soles of the skis, impeding further glide.

Field experiments (Louchet et al. 2013) show that a similar phenomenon may occur shortly after WL collapse (see section 4.3.2). The slab is cut vertically in order to bring to light the WL, whose material is then retrieved with a shovel. In doing so, the delicate bonds between crystals are broken, and the collected snow can flow like dry rice grains. But after a few seconds, it "clots" on the shovel, and cannot flow any more. Such a behavior can be easily explained and generalized to collapsed WLs using a theoretical model (appendix B) (Louchet 2015), the conclusions of which are twofold:

i) Snow grains of the collapsed WL go on flowing as long as the contact time with each other is short enough to prevent them from sintering (or "clotting") together, i.e. as long as the shear strain rate in the WL (i.e. the slab velocity divided by the WL thickness) is large enough.

ii) If the shear strain rate becomes less than a given threshold, the grains weld together, typically after a few seconds, resulting in *WL* clotting and interruption of the triggering process. Sintering being a thermally activated mechanism, the shear strain rate threshold increases exponentially with temperature.

The main parameters controlling the *WL* shear strain rate, and therefore the flowing vs clotting alternative, are the slope angle, the slab weight, and the *WL* thickness. For instance, steep slopes, heavy slabs, and shallow *WLs* may favor slab destabilization, but peripheral slab anchoring as stauchwalls would also have to be considered. Taking a potentially very large slab, peripheral anchoring would have a small influence, and the *WL* collapse would extend, keeping its fluid character, to a considerable slab area, favoring crown crack opening. Taking the example of a typical *WL* with a resistance to collapse of 480 MPa (determined from *PST* tests (Van Herwijnen and Heierli 2009)), a slab density of 300 kg/m^3, and a slab thickness of 40 cm, the collapsed zone becomes unstable for slope angles larger than 28° (appendix B Fig. B.3, and Louchet 2015).

Other usual values can be ascribed to these parameters, giving critical slopes ranging roughly between 30° and 45°, in agreement with the acknowledged preferred value of 30° for which avalanches are observed to be triggered.

If the slab is shallower and (or) anchored by outcrops, the conditions given above would not allow avalanche release. The skiers responsible for the collapse would only notice an audible "whumpf" and possibly feel some slab subsidence, but nothing else would happen.

5.5 Slab avalanche release in four steps

In order to summarize the above analysis, the different triggering steps will now be recalled, from basal crack initiation to crown crack opening, with a particular attention to how each step would either fail, or proceed up to the final triggering.

1. The first step is the initiation of the basal crack through a local collapse of the weak layer (*WL*). It usually results from a dynamical loading such as a skier's impulse, or from blasting from the surface. The associated elastic wave is transferred to the *WL*. In some cases, e.g. if the slab is either too thick or too stiff, the transmitted impulse would not be able to damage the *WL*. In the opposite case, the *WL* usually collapses on a scale comparable to ski length, i.e. larger than the critical Griffith's size, usually of the order of a few tens of cm. The initial crack is thus likely to expand below the slab, and the freed part of the slab may start gliding down-slope. Yet, the key dynamical property of *WLs* evidenced by the sintering test (section 4.3.2) is that in some conditions the collapsed *WL*, instead of gliding down-slope, may clot back into a solid, putting an end to the release process.

2. The collapsed zone expands, driven by the slab weight. The question is now whether or not the initial *WL* shear rate, i.e. the ratio of the initial slab flowing velocity by the *WL* thickness, is large enough to prevent clotting and keep the *WL* in a fluid state. If the slab is too light, or the slope too gentle, and the shearing part of the *WL* too thick, the flowing strain rate would be too low, and the *WL* would clot. The initial collapse and the incipient expansion would result in a simple whumpf, and probably a slight pang for the skiers. If not, we go to step **3**.

3. The "un-clotted" zone has now expanded over a large area. The slab is not bonded to the substrate any more, and its weight is hanging at its top. The bigger the slope-parallel size of the hanging slab is, the larger would be the tensile stress at the top.

44 Snow Avalanches

4. A total release of the avalanche obviously requires the opening of a crown crack, which initiates when the stress exceeds the slab resistance, preferentially at a weak point. Griffith's criterion also applies here: if the incipient crown crack size is smaller than the critical size for crown crack opening, a small stable crown crack would be visible at the top, and nothing else would happen. This is what occurs for instance when the slab is anchored on outcrops and on its banks, resulting in a rapid vanishing of the hanging load after crack initiation. This is the very last chance to avoid avalanche release. Otherwise, all conditions are met, the crown crack develops, the slab gets loosed from its anchor points, and the avalanche tumbles down. The crown crack expansion may be quite fast, essentially in cold brittle snow. In this case, the related sudden release of tensile stresses in the slab may result in an audible and sometimes impressive bang.

The succession of triggering steps are schematized in Fig. 5.6 hereafter.

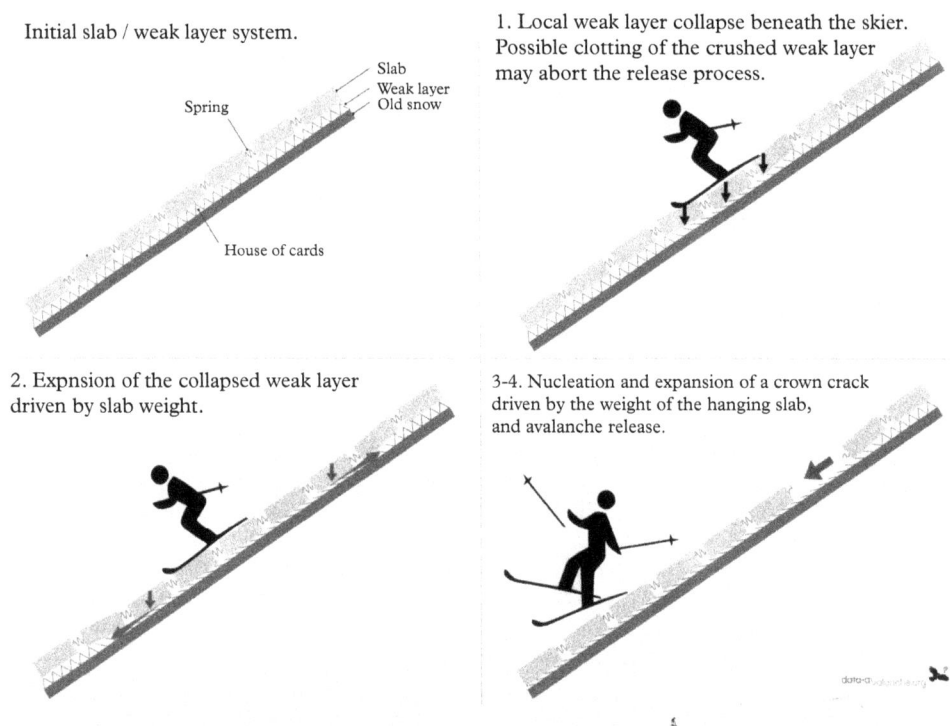

Fig. 5.6 *Schematic slab avalanche release process.* **Top left**: *Slab lying on a WL.* **Top right**: *the WL locally fails (e.g. under a skier's impulse).* **Bottom left**: *due to the slab weight, the collapsed area (basal crack) starts expanding in all directions.* **Bottom right**: *driven by the slope-parallel slab weight component, the slab starts gliding on the collapsed WL. When the weight of the suspended part of the slab reaches the slab tensile rupture stress, a crown crack opens at the top, allowing large-scale avalanching. However, at any stage, there is still one joker left, in terms of possible WL sintering, if the shear strain rate is low enough, stopping the whole process. (Louchet and Duclos 2006) (Drawing by Celine Lorentz).*

6
Superficial and Full-Depth Avalanches

Better be approximately correct than exactly wrong
Warren Buffet

The present chapter is dedicated to two geometrically extreme and in some way opposite cases of snow avalanches, respectively qualified as loose snow and full-depth ones. The former type consists of superficial flows of low-cohesion snow grains, whereas the latter essentially involves the whole snow cover that glides on the bare ground. Very few fatalities can be ascribed to any of them, which is probably the reason why they are less well studied and understood than slab avalanches.

6.1 Loose snow avalanches

In contrast with slab avalanches involving fairly cohesive snow, the term "loose snow avalanches" *sensu stricto* refers to superficial flows of low-cohesion snow grains (i.e. unable to pack).

As they usually (but not always) start from a quasi-punctual zone, they are often named "point-release" avalanches (Fig. 6.1). The smallest ones are known as "sluffs". They are often released on high-angle slopes, gradually gathering new snow grains as they propagate in fan-like shapes (Figs 6.1 and 6.2). The flow thickness increases with travel distance, due to accumulation of both superficial snow and to some extent from deeper snow layers driven by superficial flow.

Such characteristics are common with some dry sand avalanches, as illustrated in chapter 1, Fig. 1.2.

Conditions for loose snow avalanche release are often found at high elevations where fresh or faceted snow remains cold and dry, and where steep slopes help destabilization of new grains, typically more than 35° or 40°.

When triggered artificially, and in contrast with slab avalanches that break some distance above skiers, they usually start beneath them, and sometimes along ski trails, with less punctual starting zones in this case (Fig. 6.2). As a consequence, they are responsible for very few fatalities, at least for people at the origin of triggering. But larger flows may catch people standing down-slope and carry them down to gullies or cliffs. They may also trigger, in turn, slab avalanches (Fig. 6.1).

Another type of loose snow avalanche, qualified as "wet" or "moist", is also encountered (http://avalanche.state.co.us/forecasts/help/avalanche-problems/loose-wet/). Low cohesion is obtained in this case by air warming, rain, or solar radiation, transforming old or more recent surface snow into some kind of slush. Such avalanches are also usually characterized by a small or quasi-punctual starting zone, and a fan-shaped propagation. Flow thickening is usually larger, owing to gradual sintering of warm snow grains. The avalanche front stops on gentler slopes, where the decreasing shear strain rate allows global sintering (see section 5.4 and appendix B). The arrest area displays accumulation of pinwheels and rollerballs (Fig. 6.3.). In some cases, if the snow cover is wet

Fig. 6.1 *Point-release sluff, triggering a small slab avalanche further down. Pointe Longe Côte, Haute Maurienne (Savoie), 22 May 2014 13:30. (Photograph by Alain Duclos).*

Fig. 6.2 *Series of sluffs released along a ski track, Pointe d'Andagne, Haute Maurienne (Savoie), 2 April 2010, 11:00. (Photograph by Alain Duclos, Data Avalanche).*

enough, wet loose snow avalanches may drive snow down from the very bottom of the snowpack, gradually transforming into the full-depth avalanches described in section 6.2.

In both cases, the triggering and flow processes, typical of non-cohesive grains, are reminiscent of Bak's sand pile model that introduced the concept of Self Organized Criticality (SOC) (appendix A). A power-law (i.e. scale invariant) statistical distribution of loose snow avalanche sizes is

Fig. 6.3 *Rollerballs in point-release (left) and slab (right) moist snow avalanches. Haute Tarentaise, Les Arcs district, 27 April 2019, 12:30. (Photograph by Arthur Fernandez).*

therefore expected. However, to the author's knowledge, such data are still unavailable for loose snow avalanches.

A comparison of the corresponding critical exponent to the value measured for slab avalanches (chapter 4) should nevertheless be of interest. Slab avalanche release involves indeed both basal cracks and crown failures, whereas loose snow avalanches remain usually superficial (no basal crack) and start from a single point (no crown). This was the reason why the slab avalanche model by Faillettaz et al. (2004, 2006) was a two-threshold model, quite different from Bak's that seems to be more relevant in the loose snow case.

6.2 Full-depth avalanches

As mentioned in the introduction, three main types of snow avalanches are usually distinguished: slab avalanches triggered by failure of an underlying weak layer; loose snow avalanches (discussed above), triggered by the destabilization of a few snow grains that gradually knock out other ones; and full-depth avalanches, encompassing all other types together. For this reason, a variety of full-depth avalanche definitions are found in the literature, which may bring with them some confusion as to the underlying triggering mechanisms.

48 *Snow Avalanches*

Full-depth avalanches differ from slab avalanches in both release and flow characteristics, and also in preferential slope orientation, season, etc. They are still poorly known, presumably because they are responsible for many fewer fatalities than those involving slabs laying on weak layers. They nevertheless cause significant damage, and a better knowledge of their triggering processes and of their specific propagation mechanisms would help risk mitigation. The present section aims at understanding and modeling the physical mechanisms at the origin of their release. After a rapid overview of main full-depth avalanche characteristics, we propose a classification of the different types of snow involved in such avalanches, on the basis of the topological concept of percolation, from which we derive the corresponding triggering mechanisms. Flow and arrest processes are briefly analyzed in terms of healable granular media dynamics.

6.2.1 Observations

Full-depth avalanches are usually defined as slow flows of large snow volumes, most often of high density wet snow (around 500 kg/m^3). But they also occasionally involve dry snow (e.g. Col des Aravis, French Alps, December 2008 & 2010). In most cases, full-depth avalanches are described as avalanches gliding on bare ground (Figs 6.4, 6.5), but they may also glide on intermediate snow layers in the absence of any weak layer (on which the glide velocity would have been significantly larger). Their widths are usually narrower than those of slab avalanches that may cover a whole hillside. In the northern hemisphere, they often occur in November–December and in March–April, and are less frequent in January–February. (e.g. winters 2011–2012, 2008–2009, 2007–2008 in the Alps) (Coubat and Duclos, pers. comm.). As their common feature is the absence of a weak layer, their release is always spontaneous, i.e. full-depth avalanches are not triggered by skiers or explosives. Starting zone slopes are always steeper than 30°, which means in particular that triggering from horizontal terrain is impossible, also related to the absence of a weak layer. By contrast with slab avalanches, the fact that full-depth avalanches are less frequent in the depths of winter, and that they usually occur on sunny slopes, suggest that either melt or rain water plays a significant

Fig. 6.4 *Full-depth avalanche, Râteau d'Aussois, les Côtes (Savoie), 8th December 2012. (Photograph by Alain Duclos).*

Fig. 6.5 *Typical full-depth avalanche, Bessans (Savoie), 13 January 2018. (Photograph by Gregory Coubat).*

role. In addition, a strong correlation is observed between full-depth avalanche frequency and a temperature close to zero at the ground level, independently of the air temperature during the days before the release (Coubat and Duclos, pers. comm.). The uphill part of the snow layer starts gliding slowly on the ground surface, whereas the stauchwall may remain immobile, at least for a while, sometimes resulting in spectacular buckling events favored by warm temperatures (Fig. 6.6).

6.2.2 Different kinds of snow involved in full-depth avalanches

Snow involved in full-depth avalanches is usually classified as dry or wet (or damp). A more detailed distinction is sometimes made between wet in the one hand, and sodden, soggy, or waterlogged in

Fig. 6.6 *Snow buckling in a full-depth avalanche. (Photograph by Jérôme Huet).*

the other hand, despite the fact that such qualifiers are not precisely defined in the literature. We propose here definitions of damp snow varieties that will be shown in the following to correspond to quite different possible triggering mechanisms. They are based on the topological concept of percolation, defined and discussed in chapter 2.

6.2.2.1 *Dry dense snow*

Dry dense snow results naturally from gradual compaction of light snow, driven by a reduction of ice surface energy, and realized through diffusion of individual water molecules on ice grain surfaces at relatively mild temperatures. Such a process results in a thickening of ice bridges between snow grains, reducing the proportion of air vs ice volume, which significantly increases snow density, and also snow strength, by 3 orders of magnitude if density is increased from 200 to 600 kg/m^3 (Shapiro et al. 1997). A fairly similar result is also obtained by mechanical compaction, for instance during ski track grooming. This dry dense snow will be labeled "dry snow" hereafter.

6.2.2.2 *Wet snow*

Wet dense snow is formed either by bulk snow thawing at warm temperatures, by rain water infiltration, or both of them. Damp snow in which air has been replaced by water in part or in whole is therefore denser than the dry snow it comes from. In contrast with dry snow (section 6.2.2.1), bonds between snow grains have no reason to get stronger. Thawing indeed decreases the snow volume fraction, in favor of water, resulting in reduced snow strength. At a given stage of water infiltration, the water volume fraction may exceed the bi-percolation threshold (chapter 2). Snow still percolates through the snow layer, providing mechanical strength to the system, but liquid water also percolates, i.e. is able to find continuous paths through the snow layer, reach the bare ground, and lubricate the potential glide surface for the avalanche. In the following, this type of snow will be called "wet snow".

6.2.2.3 *Soggy snow*

An extreme case is that at which water still percolates, but snow grains do not percolate any more, due to an excess of water. At this stage, damp snow suddenly experiences a strong strength discontinuity that changes the medium from a solid state to a granular slurry, due to the fact that the continuous solid skeleton previously made of percolating snow has now disappeared. This is the "Turkish coffee" case mentioned in section 2.5. Such a state will be called "soggy snow".

6.2.3 Release mechanisms

6.2.3.1 Basal crack destabilization

A general feature of full-depth avalanches is the absence of a weak layer, which eliminates any basal crack initiation by collapse. Artificial triggering (human or by explosives) is therefore almost impossible. In addition, another type of glide surface for the avalanche has to be found in replacement of the weak layer. In the case of slab avalanches, the collapsed weak layer provides indeed a smooth glide plane whose slide resistance is quite low, due to the loose granular character of the crunched weak layer material. By contrast, in the present case, as the weak layer does not exist, the most frequent and obvious glide surface is the ground, if soaked by a sufficient amount of water. This situation may be achieved in different ways, according whether snow is wet, soggy, or dry, as discussed in sections 6.2.2.1 to 6.2.2.3.

Another (and fundamental) consequence of the absence of a weak layer is that we are facing a much simpler problem than in the slab avalanche case, since the only loading component involved in triggering is the shear component parallel to the slope. As a result, the basal crack critical size for unstable crack expansion in Griffith's sense (Griffith 1920), now in pure shear, is significantly larger than computed for mixed shear-compression mode in (Heierli et al. 2008a, 2008b), typically several tens to a few hundred meters instead of a few decimeters.

The subsequent triggering mechanism is therefore very simple, similar to what was described in early pure-shear models (Louchet 2000, 2001). Griffith's criterion may again be applied (but involving only a pure-shear stress component), in order to determine the critical size at which such flaws become unstable. The shear stress at the interface is given by:

$$\tau = \rho g h \sin\alpha = \rho g \left(h_{//} \cos\alpha\right) \sin\alpha$$
$$= \frac{\rho g h_{//}}{2} \sin 2\alpha \qquad (6.1)$$

where ρ is the snow density, g the gravity constant on Earth, $h_{//}$ the snow depth measured vertically, h the snow depth measured perpendicular to the ground, and α the slope angle (see Fig. B.2). Assuming that snow falls vertically on average, this shear stress goes through a maximum for a slope angle of 45°, which is a compromise between a steep slope that increases the shear component of the snow weight along the slope (proportional to $\sin\alpha$), and a gentler slope that favors a thicker deposited snow layer (proportional to $\cos\alpha$).

In such conditions, Griffith's criterion writes:

$$\tau \sqrt{\pi a_c} = K_{IIc} \qquad (6.2)$$

where K_{IIc} is the snow shear fracture toughness, and a_c is the critical flaw size. Using eq. (6.1), eq. (6.2) becomes:

$$\frac{\rho g h_{//}}{2} \sin 2\alpha \sqrt{\pi a_c} = K_{IIc} \qquad (6.3)$$

or equivalently:

$$a_c = \frac{4 K_{IIc}^2}{\pi (\rho g h_{//})^2 \sin^2 2\alpha} \tag{6.4}$$

As intuitively expected, eq. (6.4) shows that, during water percolation through the snow cover, the critical flaw size that destabilizes the avalanche is reached all the more rapidly when the snow cover is dense (ρ) and thick ($h_{//}$). Rain water contributes to a significant increase of snow density, thus decreasing the critical flaw size and favoring the release process. Eq. (6.4) also shows that a slope angle close to 45° (that maximizes $\sin^2 2\alpha$) favors triggering. This is derived under the simplifying assumption of a constant snow depth (measured vertically), but may vary under different snow accumulation configurations.

6.2.3.2 Wet avalanche release

We are facing here two percolation problems. The first one is a three-dimensional percolation through the snow layer, and the second is a two-dimensional one at the snow/ground interface (chapter 2).

As mentioned above, in the snow layer itself, snow percolates (see definition of "wet snow" in section 6.2.2.2), still providing some solid strength to the system, but water also percolates, and is drained through the snow cover down to the ground, favoring snow glide.

The presence of water at the ground level, and a temperature close to zero Celsius, are reminiscent of glacier glide kinetics. Glaciers whose temperature at ground level is negative, called cold glaciers, glide on the glacier bed at low velocities, due to significant friction. By contrast, in warmer glaciers, the temperature at the interface may reach 0 °C, resulting in ice melting. The presence of water lubricates glacier glide, significantly increasing glide velocity, and leading in many cases to instabilities and ice falls (Faillettaz et al. 2012).

The situation in the case of wet avalanches is somewhat similar. Water percolation down to the ground gradually increases its temperature, which eventually reaches the melting point at ground level. A similar effect may occur due to geothermal flux, without any contribution from water percolation. As a consequence, water accumulates at places in the form of molten snow flaws, which further develop, as discussed hereafter.

At this stage, two different phenomena may now help destabilization of the snow cover, depending on the degree of water percolation at the snow/ground interface, as shown by Faillettaz et al. (2015) in the case of glaciers.

i) During the early development of molten flaws, the snow layer remains anchored to the ground by a continuous array of frozen snow/ground interfaces, separating isolated molten zones: frozen interfaces percolate at the interface, but molten zones remain isolated from one another. Due to a "water column" effect (height of water drainage paths through the snow cover), molten flaws are pressurized, which helps in uplifting and uncoupling the snow cover from the ground.

ii) In a second stage, since bi-percolation cannot exist in two dimensions (chapter 2), if water continues being drained through the snow layer, molten zones grow, and a percolation inversion may take place. Frozen zones now become isolated and separated by a continuous array of molten ones. This inversion allows circulation and evacuation of water at the snow/ground interface. Such a mechanism results in some depressurization of molten flaws, but enhances drainage of warmer water through the snow cover. Thus, anchoring zones made of frozen spots readily disappear, considerably increasing lubrication at the bed level.

In addition to this thermodynamical effect, a mechanical phenomenon helps the expansion of molten zones. Such zones behave actually as basal cracks, experiencing the shear stress due to the slope-parallel component of the slab weight. When a molten zone gradually grows due to water circulation, it eventually reaches Griffith's critical size in shear mode, becomes unstable, and expands readily to the whole slope. Buckling of the snow cover also participates in flaw growth. The snow cover is now able to glide down-slope.

Two additional conditions must, however, been fulfilled in order to allow long-range avalanching: i) opening of a crown crack at the top of the snow layer, and ii) failure of the stauchwall at the bottom.

Owing to the large amount of water in the snow layer, the tensile rupture stress is low. Crown cracks can open easily as gaps at any weak place, in excellent agreement with observations (Figs 6.7, 6.8). The shear rupture stress being also reduced in the same proportion, snow may fail by shear from both tips of the crown crack (Fig. 6.9), resulting in several narrow avalanches being released on the same slope. The snow yield stress being also low, the plastic zone size is large, and crack tips can blunt quite easily (Figs 6.7, 6.8). The system now follows the curve in Fig. 3.4 c instead of the dashed line. Rupture becomes ductile and may be delayed, which means that it may occur only for a significantly larger deformation, or may not occur at all, again in agreement with observations (Figs 6.7, 6.8).

Fig. 6.7 *Incipient wet snow full-depth avalanche on a barn roof. Numerous crown cracks open at every weak place, due to low rupture stresses of wet snow. Note the limited lateral extension of such cracks, related to low yield stresses that favor easy blunting of crack tips (see text). Some of them have merged (on the right) giving large cracks that may trigger an avalanche. (Photograph by François Louchet).*

54 *Snow Avalanches*

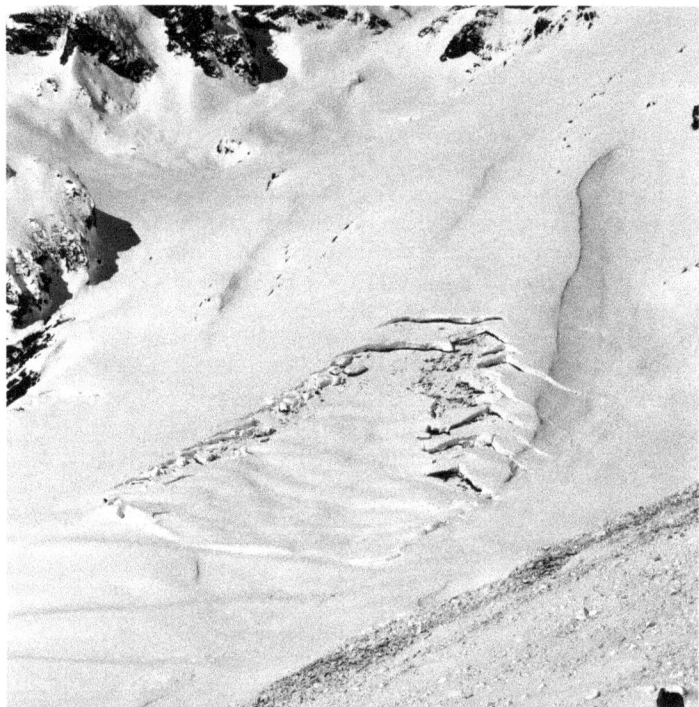

Fig. 6.8 *Incipient wet snow full-depth avalanche, Bonneval sur Arc, Savoie, France, 2018. Same comments as for Fig. 6.7. (Photograph by François Louchet).*

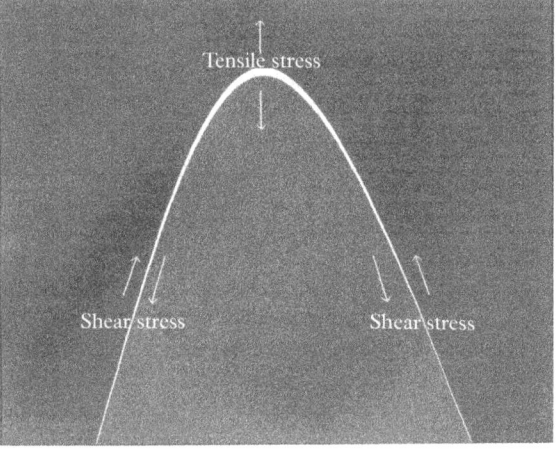

Fig. 6.9 *Idealized schematic of a tensile failure at the crown, and of lateral failures in shear mode on both sides.*

In the case of "classical" slab avalanches, the slab stiffness prevents extensive glide as long as the stauchwall does not fail. In the present case, owing to the ductile deformability of wet snow, the uphill part of the snow layer may start deforming by creep, and also sliding down-slope, whereas the downhill part may not, due for instance to an increased friction or a stauchwall. Such a phenomenon (Bartelt et al. 2012) may result in buckling events (Fig. 6.6).

6.2.3.3 Soggy snow avalanche release

At this stage, snow grains do not percolate any more. We are dealing with some kind of high viscosity granular material that may deform and flow very slowly at least in the beginning. During such a slow motion, some snow grains carried along in the flow may weld together, while bonds between other ones may break. The global slurry behavior results from a kinetic balance between these two opposite processes. The evolution equation derived in the case of collapsed weak layers in appendix B and in (Louchet 2015) can be applied here, showing that if the slowly increasing creep rate reaches a given threshold, the snow would suffer a sudden viscosity decrease (bifurcation), essentially in the deepest layers that experience larger shear stresses, and flow downward more rapidly. Such a mechanism is not incompatible with the presence of blocks of solid snow, often observed in the avalanche flow, provided they are imbedded in the slurry. This type of triggering does not involve any crown crack, at least in its traditional definition, as the slurry is already fluid.

6.2.3.4 Dry snow full-depth avalanches

The most obvious origin of water at the bottom of a dry snow layer is local thawing due to the geothermal flux from a warmer ground (otherwise the snow would not be dry!). The mechanism is similar to that of the wet snow case: continued thawing at the bottom of the snow cover gradually increases the sizes of molten flaws, which merge and eventually fulfill Griffith's criterion.

At the same time, the slope-parallel tensile load at the upper rim of the basal crack (i.e. at the crown) raises, thus increasing the stress intensity factor $K = \sigma\sqrt{\pi c}$, where c is the crown flaw size. Dry snow being an elastic-brittle medium, the situation is fairly similar to that of classical slab avalanches, except that weak layer dynamical sintering cannot take place, as tensile crack surfaces are not in contact. The system follows the straight dashed line of Fig. 3.4, and eventually reaches the critical value, i.e. the dry snow tensile toughness. A crown crack readily opens and expands rapidly, leading to avalanche release.

6.2.4 Propagation and arrest

Due to rough glide surfaces, the release of full-depth snow avalanches readily transforms the snow layer into a heavy, tumbling granular slurry, whose bottom layers smooth out the initial roughness of the ground.

Such avalanches propagate at a fairly low velocity (10 to 40 km/h). As they plough out the major part of the snow depth on their paths, the total moving mass may be significantly larger than for slab avalanches, with impressive thicknesses and densities, and may result in huge damage (mainly on buildings) in spite of their low speeds.

The flow exhibits significant density and velocity gradients and heterogeneities, making questionable a description using classical equations of fluid mechanics. In addition, as mentioned above and in chapter 4 and appendix B in the case of weak layers, the flowing snow propensity to heal (dynamical sintering) is likely to play a key role in both the propagation kinetics and the arrest mechanism of full-depth avalanches. Sintering occurs at low shear strain rates, and results in a drastic and discontinuous viscosity increase by several orders of magnitude. This is a particular case of the so-called strain-rate softening, an instability well known in materials science to be responsible for strain localization, resulting from an obvious positive feed-back. As a consequence, the lowest part of the flow, experiencing a larger strain rate, is expected to remain for a longer time in the fluid state than the upper one, ending up with a thick solid-like raft of dense snow flowing on a slippery carpet.

This phenomenon may provide a very simple alternative explanation for the impressive destructive effect of full-depth avalanches on impacted infrastructures, in spite of their low velocities. A fluid-like medium may indeed easily spread out laterally on both sides of the obstacle, while a quasi-solid one would not. The momentum transfer to the obstacle is thus significantly larger in the latter case. Classical fluid dynamics models are unable to account for such effects, except using complicated multi-parameter phenomenological models, whose physical meaning remains questionable.

6.3 Summary

By contrast with slab avalanches that are triggered by weak layer collapse and propagate on the weak layer surface, full-depth avalanches glide (by definition) on the bare ground. As a consequence, they can hardly be triggered by skiers or explosives. Water plays a key role in release mechanisms. Its origin (snow thawing at the bottom of the snow layer, bulk snow melting, or rain water drainage) determines the type of avalanche. Such types were discussed above in terms of percolation. In the case of "soggy snow", water percolates, but snow does not any more, which results in flow of a snow-water slurry. In the case of "wet snow", water/snow bi-percolation in the snow layer bulk leads to development of flaws of molten snow at the ground level. Such flaws may in turn percolate, and help avalanche release. Dry snow avalanche release is more similar to that of slab avalanches, with the key difference that weak layers do not exist in this case. In all cases, triggering is controlled by shear stresses. Flow and arrest characteristics are essentially ruled by dynamics of healable granular media, and in particular by a fluid-solid transition below a critical shear strain rate.

7
Snow and Avalanches in a Climate Warming Context

It is difficult to make predictions, especially about the future
This quote is alternatively attributed to Confucius, Mark Twain, Woody Allen, and others, *suggesting that the opposite assertion may also be true...*

As for the future, the main question is not to predict it, but to make it sustainable
Antoine de Saint Exupéry, "Citadelle", published in 1948

The influence of climate warming becomes every day more and more visible in our environment, and more particularly in mountainous areas, to such an extent that one may wonder how and how much it may affect snow cover extent and snow stability in the next future. Glacier flows are speeding up, increasing ice-fall frequency, glacier fronts are retreating more and more rapidly, permafrost thawing results in rock-falls even in north faces, making trekking and climbing activities more hazardous. In such a context, the influence of climate change on avalanching and avalanche hazard remains an open issue.

What we actually know is that the Earth is indeed undergoing an unprecedented climate change, in terms of abruptness of both greenhouse gases input rate and temperature increase. What we would like to understand and possibly predict is i) what will be the next equilibrium temperature, and ii) what would be the consequences on avalanche processes, risks, and possible mitigation procedures.

7.1 Climate change

A key point is trying to understand the mechanisms ruling climate change and climate change rate. Current climate forecasting techniques are usually based on <u>continuous</u> extrapolations of present trends, using both physical laws and a large number of parameters tuned in order to fit past climatic evolutions. Calculations are carried out starting from more or less optimistic scenarios, in order to evidence possible consequences of less or more stringent political decisions (see IPCC reports at https://www.ipcc.ch/sr15/).

At the moment, climate warming predictions are all the more worrying in that they are recurrently and more and more frequently revised upwards. The reasons for such a bias are twofold: i) simulations cannot take into account feedbacks due to forcing at a rate that has no equivalent in the past and whose consequences are inevitably unknown; and ii) extrapolations are based on the implicit assumption that a continuous cause has necessarily continuous consequences. This last point is crucial, as numerous physical processes may respond in a discontinuous manner to continuous "loadings", resulting in tipping of the system into a new equilibrium. This is particularly the case in so-called "complex systems", made of a large number of interacting elements, whose behavior is quite different from that of the addition of individual entities ("sheep herd effect"), as discussed in appendix A. Several examples are given in the present book, as Griffith criterion, slab avalanche or sluff spontaneous release, etc. As mentioned in appendix A, the tipping process would be irreversible, at least at human timescales in the case of climate.

Are these perspectives relevant for present climate evolution? The approach of a tipping point is usually announced by fluctuations of increasing amplitudes. This is exactly what is currently observed, in the form of extreme heat waves or cold spells, droughts, hurricanes, forest fires, unusual rain or snow falls, etc. This clearly means that continuous extrapolations of past climatic changes in order to predict future evolutions are just intrinsically unreliable. We are actually facing an impending climate tipping (Louchet 2016) that will bring us into an unknown state (appendix A, Fig. A.1). This assertion is confirmed by two recent papers (Steffen et al. 2018, Neukom et al. 2019). The latter evidences an "aberrant" and unprecedented global synchrony of present day warming, in contrast with the "little ice ages" of the sixteenth, seventeenth, and eighteenth centuries, that successively occurred in different parts of the planet. This global synchrony is typical of a critical point (appendix A) at which the correlation length becomes of the order of the system size.

In the case of glacier flow stability, equations based on the theory of critical phenomena (Faillettaz et al. 2015) allow quite accurate predictions of ice-fall dates, and might be used for climate tipping, provided reliable data on instabilities are available. However, the obvious and impressive increase of climatic fluctuations suggests that the tipping point is already very close, which may explain why every next continuous extrapolation of temperature rise yields more and more alarming predictions.

In such a situation, the only thing we can predict with certainty about the next stable state is that it will be hotter. The amplitude of the temperature jump may be tentatively guessed by comparison with the PETM (Paleocene-Eocene Thermal Maximum), which occurred 56 million years ago (Wikipedia. The Paleocene–Eocene Thermal Maximum (PETM), alternatively "Eocene thermal maximum 1", https://en.wikipedia.org/wiki/Paleocene-Eocene_Thermal_Maximum). At this time, considerable volcanic activity resulted in an injection of greenhouse gases in the atmosphere comparable to the present one (although for different reasons). Though spread over 20 000 years, this resulted in a global warming estimated at around 5–8 °C. The impressive abruptness of the present greenhouse gases increase (150 years only), much shorter than "geological time", may not allow a gradual adaptation of the system, and more especially of the biosphere, that might have hindered positive feedbacks. It suggests a stronger temperature response, that might be definitely larger than the 2 °C or 3 °C within which we are unsuccessfully "trying" to limit the present evolution. Even more worrying, current fluctuations suggest that such a perspective is more likely to happen during the next few years than by the end of the century. In this context, a global warming by 8 °C to 10 °C in the few next years would not seem unlikely. It is worth noting that this is a global average, from which local climate evolutions might depart, in both directions.

Another phenomenon may interfere in the future. A remarkable peculiarity of Earth is the presence of water under its three physical states, gas, liquid, and solid, with a proportion depending on latitude and altitude. Water vapor is in equilibrium with liquid water, and has therefore a self-stabilizing influence. Sea water is also in contact with ice shelves in the Arctic Ocean, and with calving glaciers in Antarctica and Greenland, keeping it locally around 0 °C. This cold water is subsequently distributed worldwide through the oceanic current network, slowing down the ocean warming rate. However, the Arctic ice shelf tends to shrink more summer after summer, and the retreat of Antarctic and Greenland glaciers gradually leads to a situation where they will eventually stop being in contact (and therefore in thermal equilibrium) with sea water, allowing enhanced ocean and atmosphere warming rates. In addition, water vapor is the dominant greenhouse gas in our atmosphere. So far, owing to vapor saturation, additional water vapor input at constant temperature was balanced by an equivalent amount of precipitation in the form of rain or snow. But water vapor saturation content increases with temperature, which may result in a positive feedback loop, i.e. an autocatalytic effect on climate warming.

Such perspectives are all the more alarming given that the present climate change is taking place over a dramatically short period. This situation may endanger humankind's survival, and absolutely requires both an exceptional and immediate effort to reduce greenhouse gas concentrations, and a powerful taking off in resilience strategies.

7.2 Possible consequences for avalanching

Which consequences should we expect for snow and avalanches?

A few studies dealing with this topic have been published during the last 20 years. They are usually based (again) on the implicit assumption that climate change processes are continuous and that future evolutions may be extrapolated from past records, which is quite unjustified as shown above. Some of them point at an increase of avalanche frequency and run-out distances, evidenced for instance in the Indian Himalayas through tree-ring based snow avalanche reconstruction (Ballesteros-Cánovas et al. 2018) or in Colorado (Berwin 2019), and considered as mere and direct consequences of climate warming. However, there is no obvious or simple reason why an increase of climatic disturbances of any kind could be ascribed only to climate warming as such. By contrast, as mentioned above and in appendix A, such an enhancement of fluctuations of all sorts of climatic-related phenomena, and in particular of snow falls or avalanching activity, should be considered as a warning sign of the proximity of a tipping point. Such an avalanching upsurge is expected to grow as long as the tipping process is not over. In such a transient period, the succession of heavy snow falls and thawing episodes would probably favor spontaneous full-depth avalanches with larger run-out distances.

On the other hand, beyond the transition, the system would reach a warmer new equilibrium state, where such bursts are expected to vanish, or at least to decrease sharply; that would be no great consolation, however. But precise avalanching activity in the new equilibrium state is all the more difficult to predict as the characteristics of the new state are totally unknown, except for a speculative and rough estimate of temperature increase.

Nevertheless, a few avalanching trends may be devised through an estimate of a future temperature vs altitude profile. The vertical atmosphere temperature gradient significantly depends on humidity, which is an additional source of uncertainty. It is larger in a dry atmosphere, of the order of 9 °C/km, but an average of 6 °C/km is usually acknowledged.

Taking a warming of 6 °C, which might well be an underestimate (see section 7.1 above), with a gradient of 6 °C/km, and assuming that this gradient would not be significantly modified during climate tipping, temperatures comparable to present ones would be found 1000 m higher up. A typical "medium range" European ski resort with ski run elevations between 1700 m and 2500 m, for instance, would be forced to shift up to elevations between 2700 and 3500 m (if any are available) to survive! Many of them would disappear. Skiing would become restricted to a small number of privileged high altitude areas. Such a change would probably discourage some people from off-piste skiing and back-country activity, while groomed tracks in a few resorts may be maintained for a while in more or less acceptable conditions owing to expensive artificial snow grooming.

In addition, global warming would not change significantly the atmosphere thickness. As greenhouse gases (GHGs) essentially control infrared radiations, the intense shorter wavelength solar radiation experienced at high altitudes should not be affected by warming. Such conditions are expected to favor formation of near-surface large faceted crystals resulting in snowpack instability, as already observed in high altitude tropical mountains (Williams et al. 2001). On the other hand, enhanced glacier flow speeds and associated retreat, as well as speeded up permafrost thawing, would increase the frequency of both ice- and rock-falls, that may in turn trigger avalanches, threatening neighboring inhabited zones if any.

Unfortunately, in such a climate change context, avalanche-related fatalities in concerned countries should be put in perspective with considerably more dramatic worldwide consequences resulting from expected floods, droughts, agricultural collapse, people migrations, etc.

8
Summary and Conclusion

*I have noticed that even people who claim everything is predetermined
and that we can do nothing to change it, look both ways before they cross the road*

Stephen Hawking

*As the buoys marking the shoals are often out of position,
mariners are cautioned to be on their guard when navigating these zones*

Gerald Durrell, *"My family and other animals"*, Penguin Books, 1959

From the most remote times of humanity, frightening natural phenomena were thought to arise from divine decisions, thrown down onto humans in punishment of supposed mischief. This is the case for avalanches, and the reason why lots of shrines, chapels, and oratories were built in many mountainous areas such as the Himalayas or the Alps. The numerous baroque chapels in Savoie are nice and interesting examples.

Such beliefs were gradually taken over by other traditions, based on practical knowledge and popular wisdom built up by experience of successive generations. Old villages were established in "avalanche-free" zones, which means in zones where no avalanches were recorded during several centuries. Unfortunately, such a popular wisdom cannot always resist real estate pressure, resulting in building damage and human fatalities. Another belief relies on a misinterpretation of so-called recurrence times. The common opinion is indeed that if a catastrophic event like a flood or an avalanche of a given size has been observed at two times separated by e.g. 40 years, nothing similar would happen in the next 40 years. This is obviously wrong. A recurrence time of 40 years for an avalanche of a given size does not mean that the phenomenon will occur periodically every 40 years, but that it has a probability of 1/40 to occur every year, which is totally different. As mentioned in the present book, such probability estimates may be easily extended to bigger avalanches, characterized by much longer recurrence times, even exceeding human memory. The scale-invariant character of starting zones size distributions indeed obeys power-laws (straight lines in double-logarithmic plots). Normalizing the occurrence data for small avalanches on a given zone provides thereby an estimate of the recurrence time for much bigger and uncommon events by extrapolation of the occurrence frequency power law.

Another usual misunderstanding is the implicit belief that snow behaves as a traditional compact solid. Interestingly enough, it does not. The use of physical laws well established for engineering materials, such as Coulomb's friction law for instance, may be frequently misleading if applied to snow release processes, and should be in any case handled with extreme caution. On the other hand, despite the fact that avalanche flow mechanisms are in principle beyond the scope of the present book, it is worth noting that flowing snow does not behave as a conventional fluid either. It cannot be considered as Newtonian since its viscosity exhibits a sharp transition for a critical shear strain rate, strongly suggesting that the use of Navier–Stokes equations is often inadequate in this case. Flow and arrest simulations of snow flow should incorporate a shear-rate threshold in its dynamic viscosity.

For these reasons, the analysis of avalanche release mechanisms outlined in the present book is based on specific properties of snow, characteristic of its granular nature and specific topology, and in terms of its associated particular deformation, rupture, and friction properties, that may be in strong contrast with those of compact solids. It is also based on quite useful mathematical concepts, namely theories of critical phenomena, dynamical systems, bifurcations, percolation, etc.

In the field data chapter (chapter 4), we focused on statistical distributions of starting zone sizes, which obey power laws, and on specific properties of weak layers, prone to collapse under mechanical loadings, but also to occasionally sinter if their shear strain rate does not exceed a given threshold.

In terms of modeling, slab avalanche release was analyzed at two different scales, both of them in terms of critical phenomena. At a global level, the scale invariance of starting zone sizes recorded for a large number of events can be reproduced using a cellular automaton, showing a possible Self-Organized Critical (SOC) character, supported by the obvious independence of the different events' locations. At a smaller scale, the incipient destabilization of individual slabs was analyzed in terms of Griffith's rupture criterion, controlling the instability of the initial collapse. However, sintering of the crushed snow in the collapsed weak layer may be responsible for further slab slide abortion, explaining why the actual number of large-scale flows is significantly smaller than that of incipient triggering events. As for superficial flows and full-depth avalanche release, they are more reminiscent of dry sand flows for the former, and pure shear destabilization for the latter.

Except for artificially (and more specifically human) triggered avalanches, release prediction methods are essentially based on critical phenomena concepts. The time evolution of a system towards failure obeys general laws, typical of such phenomena, and characterized by fluctuations of growing amplitudes that obey power-laws and diverge at the critical point. Such features theoretically allow failure time prediction at least for self-evolving systems. Glacier breakdown, for instance, can now be predicted with an impressive accuracy, introducing beacon displacements in equations describing the approach of a critical point.

However, snow is not thermodynamically stable, much less so than glacier ice. Applying this technique to snow, which suffers large and fast spatio-temporal variability, is all the more difficult in that monitoring a large number of continuously evolving slopes would require significant effort and investment. Nevertheless, it may be worth testing the use of simple optical displacement beacons in order to predict some natural avalanches in particularly hazardous slopes. Future developments in the artificial intelligence field may probably be also of interest.

By contrast, the fast, unpredictable, and decisive action of a skier cannot be taken into account and introduced in advance in a computer, making accurate and deterministic artificial avalanche prediction intrinsically impossible by this type of technique.

Nevertheless, the series of successive physical mechanisms responsible for slab avalanche triggering is now perfectly known, providing a precise understanding of the involved physical processes. This should help ski resort managers determine more personalized and accurate danger scales, and practitioners facing unexpected situations make rapid and wise decisions. Additional information and recommendations on risk management for practitioners may be found on the Data-avalanche site, more precisely at: http://www.data-avalanche.org/understand., and in the excellent book by Bruce Tremper (2008).

Finally, a perspective on the future of snow and avalanche characteristics was tentatively explored in the context of the unprecedented and probably irreversible present climate warming.

Appendix A
Complexity and Critical Phenomena

> *When we try to pick out anything by itself,*
> *we find it hitched to everything else in the Universe*
>
> John Muir

A.1 From simple to complex systems

A frequent belief in physics (and in science in general) is that every event results from a single cause. This is nothing but a helpful assumption quite often used to solve rather "simple" problems, in which a main cause is isolated from a number of other parameters considered of minor importance and neglected for the sake of simplicity. Considering Newton's famous apple story (or legend), the only interaction he took into account was Earth's attraction on a single fruit. Gravitational interactions with other apples, shivering leaves in the apple tree, running beetles in the grass, or the debates of his colleague Members of Parliament at the House of Commons were neglected.

On the opposite hand, such a simplifying assumption is no more valid in complex systems, defined as assemblies of large numbers of interacting entities. The consequences of a series of chain reactions are all the more unpredictable as minor but unavoidable uncertainties in initial conditions may significantly affect final predictions.

Let us consider a set of dominos laid out vertically and at random on a horizontal plane. What happens when a domino is knocked over? The precise final result obviously depends on the exact position of each domino and initial impulse given to the first one, which is never known with an infinite accuracy. This is the well-known "domino effect" (or "butterfly effect"). This is why the problem is often addressed using different methods based on the theory of complex systems.

If the domino average spacing is large enough, the destabilized domino may fall down alone, or possibly bring down only a couple of other ones. For a tighter average spacing, the first fall will be transmitted to second and third neighbors that would fall down in turn. If dominos are laid out at random, the larger the average domino density (number of dominos per square meter), the bigger would be the "avalanche" sizes. It can be easily imagined that there exists a well-defined domino "density" for which a single avalanche may reach infinity (in fact the size of the system). This is called the critical point.

Such a chain reaction divergence is well known in the nuclear field. A nuclear power reactor in principle operates in a sub-critical regime, for which the fission of a Uranium nucleus triggers on average that of less than another one. But when fission of the considered nucleus leads to fission of more than another one, the reaction diverges, resulting in a nuclear explosion. These two regimes are separated by a critical point, for which a fission event yields exactly another one. Operation of "control rods" allows control of the multiplication rate of neutrons, responsible for nuclei fission. A reactor that would have gone beyond the critical point cannot step backwards, as the explosion takes place after quite a short delay. It is clear that a continuous variation of the control rods level may result in a discontinuous effect.

Examples of overcoming a critical point are numerous in nature and elsewhere: bursting propagation of epidemics, extinction of vegetal or animal species or civilizations, climate tipping, collapses of economic or political systems, or more prosaically rupture of solids. All of them result from continuous variations of "control parameters", and lead to discontinuous and often irreversible consequences.

The way in which a critical point may be reached and overcome is schematically illustrated in Fig. A.1, taking the example of a possible climatic tipping. We are interested in how a "control parameter" (Green

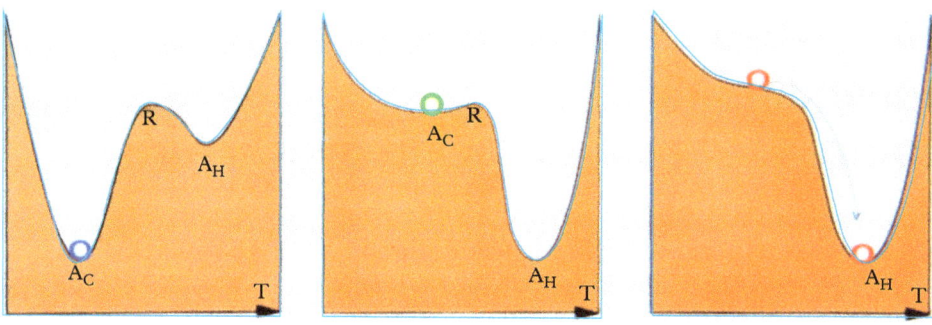

Fig. A.1 *Approach and overcoming of a tipping point. Flattening of the potential profile in the vicinity of the tipping point (middle figure) favors enhanced fluctuations of the order parameter T, and of other related quantities (see text).*

House Gases concentration in this case) may trigger a discontinuous change in a so-called "order parameter" (here the atmosphere temperature). The system stability is schematized by a potential profile. Temperature increases from left to right. Thalwegs (valley bottoms) A_C and A_H are stable positions, called attractors. They are separated by a ridge R (unstable point) called a repulsor. The system initially lies in the cold attractor A_C (left). As external conditions change (here Green House Gases concentration), the valley profile changes (middle). A_C and R eventually merge, and the system jumps in a discontinuous way into the next (hot) attractor A_H. The process is not directly reversible: if the profile is brought back to its initial shape, the system would nevertheless remain stuck in A_H. Tipping the system back to the initial stable state A_C would require a further deformation of the profile to a point where A_H and R would merge in turn.

An interesting point is that the A_C thalweg shape flattens as the critical point is approached. The system usually experiences fluctuations around the stable point, due to parameters not taken into account in the potential profile (e.g. a butterfly flapping its wings in a small island somewhere in the middle of the Pacific Ocean). As the thalweg flattens, the driving force that recurrently brings the system back to the equilibrium point A_C would gradually decrease, resulting in fluctuations of growing amplitudes, and eventually vanish at the tipping (or critical) point. In the case of climate, such fluctuations (heat or cold waves, hurricanes, heavy rain or snow falls, etc.) are a warning sign of an impending tipping, and should be seriously taken into account. In the case of glaciers, fluctuations of ice flow velocities are successfully used to determine icefall dates (Faillettaz et al. 2012, 2015).

The concept of criticality, and the specific features associated with the approach of a critical point, are indeed of considerable interest in the failure of inhomogeneous solids. The flaw destabilization described by Griffith's criterion in section 3.3 is a critical point. If applied to an inhomogeneous material, such as a rough fabric for instance, final flaw destabilization is announced by a series of fiber ruptures of increasing amplitudes. Glaciologists have recently taken advantage of this phenomenon, known as "critical slowing down", to predict with an impressive accuracy (about 48h) glacier breakdown events. Evolution of glacier flow velocity oscillations are carefully recorded using GPS beacons. Recorded data are analyzed in the framework of the theory of critical transitions, allowing in-time evacuation of endangered populations (Faillettaz et al. 2015). Snow is another example of an inhomogeneous material, and destabilization of snow layers should exhibit such premonitory events. Despite the fact that artificially triggered slab avalanches (including human-triggered ones) cannot exhibit by definition any premonitory signal, application of the technique mentioned above to spontaneous slab or full-depth avalanches should be contemplated. Nevertheless, owing to the large sensitivity of snow to atmospheric conditions (section 2.3) and to challenging technical requirements, they are not directly applicable at the moment. Further progress in this field would be of considerable interest.

A.2 Scale invariance and self-organized criticality

A further insight into critical phenomena is illustrated by sand piles (Fig. A.2), as made famous by the well-known cellular automaton (see chapter 4) devised by Bak et al. (1987, 1988), that introduced the concept of "Self Organized Criticality" (SOC). When the top of the pile is fed by pouring sand in a slow and continuous manner, avalanches are spontaneously and successively triggered. Each avalanche is a chain reaction of tumbling sand grains resulting from a bifurcation as in Griffith's case (chapter 3). However, Griffith's approach deals with homogeneous solids, whereas Bak et al.'s simulation generalizes the problem to heterogeneous and disordered media. And this is the point where collective and statistical aspects arise, requiring a statistical treatment.

In addition, feeding continuously the sand pile, the slope gradually steepens, and the avalanche frequency and size increase with the slope angle. The system eventually "diverges" at the critical slope, at which avalanches of any size, randomly triggered on the whole system, "self-organize" and stabilize the slope around a constant value in average. A quite interesting point is that, if the loading rate is low enough, each avalanche stops before the next one is released, i.e. there is no overlapping between them. In this case, the statistical distribution of avalanche sizes displays a remarkable feature: the number of avalanches $N(L)$ of a given size (width or length) L obeys a power law function of their size:

$$N(L) \propto L^{-b} \qquad (A1)$$

where b is positive. In other words, in a double logarithmic plot, the distribution is represented by a straight line with a negative slope. This means of course that bigger events are less likely to happen than smaller ones (Fig. A.3), but in a well-defined proportion. Such a feature is called "scale invariance".

This concept of "scale invariance" is fundamental in the study of critical phenomena (Sornette 2000). A power law function $y(x)$ (like $N(L)$ in eq. A1) is "scale invariant" since:

$$\frac{y(\lambda x)}{y(x)} = \frac{(\lambda x)^m}{x^m} = \lambda^m \qquad (A2)$$

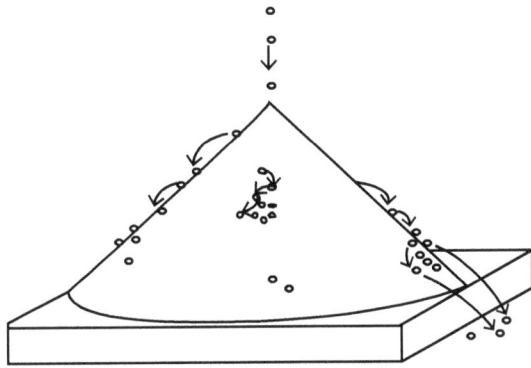

Fig. A.2 *Avalanches on a sand pile, illustrating Self Organized Criticality.*

66 Snow Avalanches

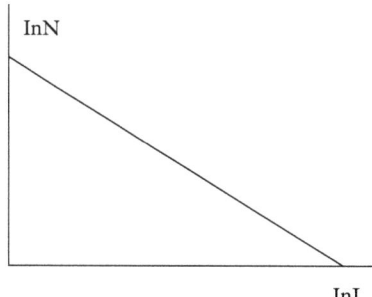

Fig. A.3 *Schematic probability density (i.e. non-cumulative power-law distribution)* $N(L) \propto L^{-b}$ *of slab avalanche widths L, giving a straight line of slope (−b) in a double logarithmic plot. Due to integration, the exponent for <u>cumulative</u> distributions of widths would be (−b+1) = −(b−1). In the same way, for the probability density of areas $A = L^2$ write* $N(A) \propto L^{-b} = (L^2)^{-b/2} = A^{-b/2}$, *with an exponent (−b/2). The corresponding cumulative distribution of areas has an exponent (−b/2 + 1) = −(b/2 − 1).*

with $\lambda^m = 1/\mu$. For a given value of the exponent m (that characterizes the studied physical phenomenon), the value of y at two different scales only depends on the ratio λ between both scales. This remarkable feature is characteristic of SOC systems.

Scale invariance, equivalent to self-similarity, is reminiscent of so-called "fractal structures", and is widely represented in nature. A well-known example is the geometry of some kinds of romanesco cabbages (Fig. A.4) in which small structures are similar to, but more numerous than, bigger ones, by a well-defined scaling factor ("fractal exponent").

Another example of SOC is that of plastically deforming crystals. Plastic deformation results from propagation of mobile crystalline defects called dislocations that experience strong elastic interactions. Application of an external load to the crystal superimposes an additional elastic field to the initial heterogeneous dislocation interaction field. As a result of this superimposition, a dislocation located at a particularly stressed point may start moving. In doing so, it modifies the elastic field experienced by nearby dislocations, which may move in turn, and so on. This situation is quite similar to the domino effect discussed above, and results in dislocation avalanches. The amplitudes of such avalanches can be measured using acoustic emission techniques for instance, and reveal that they usually obey a power law! Remarkably, this behavior is quite general, the SOC exponent being exactly the same for plastic deformation of all types of crystals with different crystalline structures (Miguel et al. 2001, Weiss et al. 2007, 2015, Weiss and Grasso 1997), and also for amorphous polymers (Deschanel et al.). The same SOC behavior (with a different exponent, however) is also found in earthquake magnitude statistics, the power law mentioned above being in this case the well-known Gutenberg–Richter law, from which the famous Richter magnitude scale has been established (Gutenberg and Richter 1956, Sornette and Sornette 1989). This similarity between earthquakes and dislocation avalanches is intuitively obvious, since both phenomena involve short strain increments connected together by elastic interactions.

In a similar way, most gravitational flows exhibit such scale invariant statistics. This is the case for snow avalanches, as shown in chapter 4, but also for rock-falls (Dusseauge et al. 2003), landslides, and turbidites (Chaytor et al. 2009). The observation of systems with comparable exponent values m suggest

Fig. A.4 *Fractal structure of a romanesco cabbage. (Photograph by François Louchet).*

that similar physical phenomena may be involved, which is quite useful for theoretical formalization tasks. They are said to belong to the same "universality class". Field data discussed in chapter 4 show that starting zone sizes obey a power law. The cellular automaton model detailed in section 5.3 gives a similar distribution law, confirming the assumptions introduced in the automaton, and suggests that a system of independent slab avalanches may be considered as a SOC system.

Appendix B
Modeling a Fluid to Solid Phase Transition in Snow Weak Layers: Application to Slab Avalanche Release

> *Mathematicians are like termites.*
> *Better immediately realize when they start being interested in your business*
>
> Claude Le Bris

B.1 A fluid to solid phase transition in healable granular materials

The sintering (or "clotting") transition reported in section 4.3.2 is explored here using a mean-field model, in the spirit of the theory of dynamical systems, analyzing the kinetics of bond failure and reconstruction. The original and full version of this general study, applicable to any kind of biphasic slurries and particularly to collapsed snow weak layers, can be found in (Louchet 2015).

We describe the slurry as a fluid medium F made of solid clusters (aggregates of clotted grains) embedded in a granular "liquid" L (free grains). We compute the rate at which solid clusters form, grow, or disaggregate during shear displacements that control contact times for both grain–cluster and grain–grain interactions. The concentrations of grains belonging respectively to clusters, granular liquid, and cluster/granular-liquid interfaces are labeled N_C, N_L, and N_I in such a way that:

$$N_L + N_c + N_i = 1 \tag{B1}$$

The system is in a pure solid state when $N_c=1$, and in a pure "liquid" state when $N_c = 0$. Topologically, in between those two extreme situations, the system can in principle be found in three different states: i) <u>fluid phase</u>, or slurry, i.e. percolating "liquid" L that may contain solid clusters; ii) <u>bi-percolated state</u>, in which both "liquid" and solid phases percolate through the system; and iii) <u>"pseudo-solid" state</u>, i.e. percolating solid phase that may contain "liquid" bubbles.

Flow in the two last states is controlled by creep of the percolating solid phase, which is of quite a different nature than that of the former one, with a flow rate orders of magnitude slower than that of the fluid state. For this reason, during the avalanche triggering period, we neglect the creep rate of states ii) and iii), that will be considered to behave as a purely elastic solid, and investigate the behavior of the system starting from the fluid (i.e. slurry) state.

In such a slurry, the solid phase is expected to percolate through the system at a well-defined threshold N_p, around 0.3 for randomly packed spheres (Powell 1979, Consiglio et al. 2003). Due to shear flow, this threshold is expected to be somewhat larger in dynamical conditions than in static ones. Nevertheless, we shall restrict the validity of the present model to N_c values smaller than 0.3.

In this domain, and due to minimization of interface energy, solid clusters should be more or less spherical, with an average radius R. Assuming that individual grains forming clusters are also spherical, with a radius r:

$$N_i \approx 4\pi R^2 / \pi r^2 = 4R^2/r^2 \quad , \quad N_c \approx (4/3)\pi R^3 / (4/3)\pi r^3 = R^3/r^3$$

and therefore N_i should be related to N_c by:

$$N_i \approx 4N_c^{2/3} \tag{B2}$$

We first study the system loaded in imposed strain rate conditions. The "liquid" phase is fed by particles taken from cluster interfaces; this process is driven by the fluid phase shear, and its kinetics is therefore assumed to be proportional to shear strain rate $\dot{\gamma}$ and to grain concentration at interfaces N_i.

On the other hand, particles are taken out from the "liquid" phase L as they form bonds with either other particles in L (cluster nucleation) or at liquid/cluster interfaces (cluster growth). Regarding bond formation between two grains in L, it is worth noting that, in contrast with gas kinetics in which the reaction rate between molecules scales as the square of their concentration, we deal here with a dense phase in which each grain of L is always in contact with other ones. As a consequence, the reaction rate is not limited by the probability of grain collisions in a dilute medium, but scales as the grain concentration N_L only (actually $N_L/2$ in order to avoid counting the same grain twice). For the same reason, the reaction rate between one grain in "liquid" L and a cluster scales as the concentration N_i of grain interfaces only. Finally, the reaction kinetics of such bond formation is taken as proportional to the contact time between "liquid" and interfaces, which obviously scales as $(1/\dot{\gamma})$.

As a consequence, the global reaction kinetics describing the evolution rate dN_C/dt of the concentration of grains belonging to clusters (hereafter labeled $\dot{N}_C$) as a function of N_C, is given by the evolution equation:

$$\dot{N}_C = -A\dot{\gamma} N_i + \frac{B}{\dot{\gamma}}\left[\frac{N_L}{2} + N_i\right] \tag{B3}$$

or, using eqs (B1) and (B2):

$$\dot{N}_C = -4A\dot{\gamma} N_C^{2/3} + \frac{B}{\dot{\gamma}}\left[\frac{1-N_C}{2} + 2N_C^{2/3}\right] \tag{B4}$$

where A is a dimensionless constant, and $B = B(T)$ (with dimension t^{-2}) is a function of temperature, that characterizes diffusion kinetics of water molecules at cluster interfaces:

$$B(T) \propto \exp\left(\frac{-Q}{kT}\right), \tag{B5}$$

Q being the activation energy for the diffusion process, and k the Boltzmann constant.

Actually, it is easier to argue in terms of imposed stress (determined by both slope angle and slab weight) than of imposed strain rate. In order to transform equation (B4) into an imposed-stress equation, we need to define a viscosity η characteristic of the fluid phase only:

$$\eta = \tau/\dot{\gamma}, \tag{B6}$$

where τ is the shear stress.

Defining a physically based viscosity of such a biphasic medium is not straightforward, as it has to take into account particle collisions, energy dissipation, etc. The viscosity of a suspension of spheres in a liquid medium, derived on such bases by Einstein in 1906 and 1911, was analyzed by Hughes (1954), and can be expressed in our notation as:

$$\eta = \eta_0 \frac{1 + N_C/2}{1 - 2N_C}, \tag{B7}$$

where η_0 is the residual viscosity of the pure "liquid" (i.e. without any solid cluster). This expression is valid only for dilute cluster suspensions, i.e. N_c close to 0, and diverges for $N_c = 1/2$. The validity of such

Appendix B: Modeling Fluid to Solid Phase Transition

a viscosity for increasing N_C values is also limited by topological constraints due to solid percolation, as mentioned above. In contrast with all previous models, a percolation-based approach of the viscosity of concentrated suspensions was proposed in (Campbell and Forgacs 1990). However, for the sake of simplicity, and because we are only interested in the evolution of the fluid phase, we shall use here Einstein's expression, which is a good approximation for N_C values smaller than the percolation threshold $N_p \approx 0.3$.

Using eqs (B6) and (B7), eq. (B4) becomes:

$$\dot{N}_C = -4A\frac{\tau}{\eta_0} N_C^{2/3} \frac{1-2N_C}{1+N_C/2} + B\frac{\eta_0}{\tau}\frac{1+N_C/2}{1-2N_C}\left[\frac{1-N_C}{2}+2N_C\right] \tag{B8}$$

In order to decide whether clusters form, grow or disaggregate, we have to know if N_C increases or decreases with time. In other words, we are interested in the sign of $\dot{N}_C$, which is the same as that of the dimensionless variable ψ defined by:

$$\psi = \dot{N}_C \frac{1}{B}\frac{\tau}{\eta_0} \tag{B9}$$

$$= -4\frac{A}{B}\left(\frac{\tau}{\eta_0}\right)^2 \dot{N}_C^{2/3} \frac{1-2N_C}{1+N_C/2} + \frac{1+N_C/2}{1-2N_C}\left[\frac{1-N_C}{2}+2N_C^{2/3}\right]$$

$$= N_C^{2/3}\left[2\frac{1+N_C/2}{1-2N_C} - \Sigma\frac{1-2N_C}{1+N_C/2}\right] + \frac{1-N_C}{2}\frac{1+N_C/2}{1-2N_C} \tag{B10}$$

where the influence of stress is represented by the dimensionless variable:

$$\Sigma = 4\frac{A}{B}\left(\frac{\tau}{\eta_0}\right)^2 \tag{B11}$$

Figure B.1 shows typical $\psi(N_C)$ curves parameterized by Σ for N_C values between 0 and 0.3, as justified above.

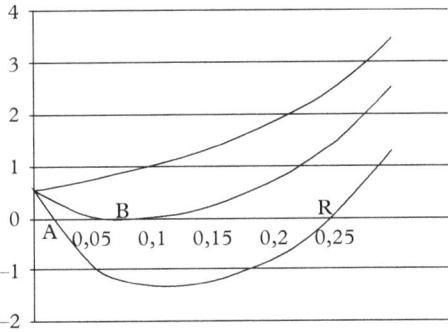

Fig. B.1 Typical $\psi(N_C)$ curves parameterized by Σ. Top: $\Sigma = 1$ (low stresses), middle: $\Sigma = \Sigma^* = 7.04$ (transition), bottom: $\Sigma = 15$ (high stresses). A and R are respectively the attractor and the repulsor of the system in the bottom curve state. They merge at the transition, yielding a bifurcation B from the fluid to the clotted state.

For $\Sigma < \Sigma^*$, i.e. low stresses and (or) large viscosities (top curve), $\psi(N_C)$ is always positive, and so is $\dot{N}_C$. Starting from any initial N_C value, N_C continuously increases and the fluid eventually clots into the solid phase.

By contrast, for $\Sigma > \Sigma^*$, i.e. large stresses and (or) small viscosities (bottom curve), the curve exhibits a negative minimum and intersects the N_C axis at A and R, that are the fixed points of the system ($\dot{N}_C = 0$), corresponding to two N_C values, N_R and N_A ($N_R > N_A$). A is an attractor, since an upward fluctuation of N_C results in a negative $\dot{N}_C$ value, that brings the N_C value back to N_A, and conversely for a downward fluctuation. A similar argument shows that R is a repulsor. As a consequence, starting from any N_C value larger than N_R drives the system towards the solid phase, whereas starting from any $N_C < N_R$ brings the system to the attractor A.

In between, both fixed points merge for $\Sigma = \Sigma^* = 7.04$ (intermediate curve), corresponding to $Nc = N_C^* \approx 0.07$, which means that the stable fluid found at $N_A < N_C^*$ contains more than 93% "liquid". This proportion increases even further with Σ, consistently with the assumption, used in equation (B7), that we are in the validity domain of Einstein's viscosity expression.

B.2 Application to slab avalanche release

Such results can now be qualitatively compared to the field observations recalled in chapter 4. Let us start from a granular fluid already flowing on a slope under its own weight, like dry rice. If the slope is steep enough, the driving force (gravitational shear stress) is large, and Σ may exceed the threshold value Σ^*. In this case, the system is trapped around the attractor A, and the fluid goes on flowing. If the slope is gradually reduced, so is the shear stress, Σ decreases, the attractor shifts towards slightly larger N_C values, but the fluid keeps flowing until Σ reaches the Σ^* value for which the two fixed points A and R merge. Beyond this point, ψ becomes positive everywhere, and the fluid clots into the solid phase S, as observed.

Let us now start from an unloaded clotted solid ($\Sigma = 0$, $N_C = 1$). We first load the system in shear, up to a stress corresponding to a non-zero Σ value. Since the system is solid, there is no flow, $\dot{\gamma} = 0$, and it cannot depart from the solid state, whatever the stress might be. However, keeping the same Σ value, we can change the initial conditions giving a mechanical shock in order to break bonds and temporarily decrease the N_C value and bring the system into the fluid domain, i.e. $N_C < N_p \approx 0.3$. If Σ is smaller than Σ^* (moderately slanted slope, top curve in Fig. B.1), ψ being always positive, the system clots back again. On the opposite hand, for Σ values larger than Σ^* (steep slope, bottom curve in Fig. B.1), two situations may be contemplated. A weak shock may temporarily bring N_C from 1 down to a value still larger than N_R. In this case, despite the large Σ value, ψ remains positive, and the system clots back again. By contrast, a stronger shock may bring N_C to a value smaller than N_R, i.e. into the attraction basin of A. In this case, the system readily becomes fluid, converges at $N = N_A$, and starts flowing down. For very large Σ values, N_R being large, even a tiny shock may destabilize the system. All these predictions qualitatively agree with observations reported in chapter 4.

We shall now explore more thoroughly the consequences of this model on basal crack nucleation and propagation at interfaces between slabs and substrates, and on the conditions favorable to avalanche triggering.

We shall use the following definitions:

h: "real" slab thickness, i.e. measured perpendicular to the slab
$h_{//}$: slab thickness measured vertically
w: thickness of the collapsible part of the WL measured perpendicular to the slab
$w_{//}$: thickness of the collapsible part of the WL measured vertically
δ: residual weak layer thickness after collapse, measured perpendicular to the slab
$w+\delta$: weak layer thickness before collapse, measured perpendicular to the slab

For the sake of simplicity, we first consider a slope with infinite size, and with a slope angle α with respect to the horizontal. Assuming no wind conditions, we consider that snow falls vertically. As a consequence,

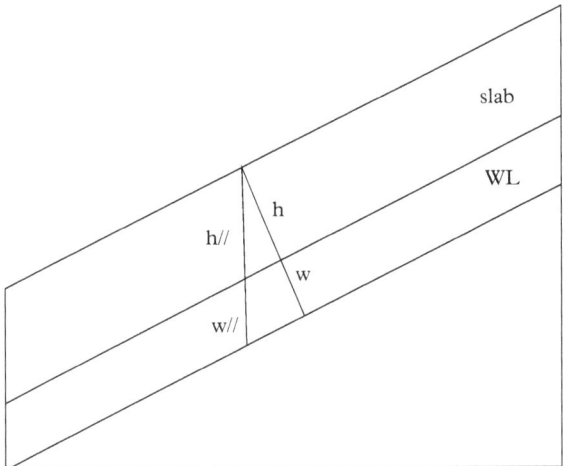

Fig. B.2 Schematic slope with a slope angle α with respect to horizontal. The thicknesses of the slab and of the weak layer (before collapse) are respectively labeled h and w if measured perpendicular to the slope, and $h_{//}$ and $w_{//}$ if measured vertically.

the resulting slab has a constant vertical thickness $h_{//}$, corresponding to a "real" thickness $h = h_{//} \cos \alpha$ measured perpendicular to the slope, that obviously decreases at constant $h_{//}$ for increasing slope angles.

WLs usually consist of either faceted grains or buried surface hoar. Faceted grains formation within the WL is a thermodynamical process, driven by temperature gradients, that does not directly depend on gravity. In a same way, surface hoar formation, that occurs during humid and clear nights, results from crystal growth at snow surface driven by humidity and temperature gradients, independently of the thickness of the snow layer on which it grows. We can therefore consider as a first approximation that, in both cases of faceted crystals or of buried surface hoar, WL thicknesses $w+\delta$ and w measured perpendicular to the slab, as well as WL mechanical properties, are constant along the slope. Avalanche release results from two successive stages, WL failure initiation, and subsequent WL shearing and downslope slide of the slab, which will be examined hereafter.

i) WL failure initiation

This first stage takes place under a combination of compression and shear load components. A WL failure can be initiated for instance by a skier, whose impulse may locally crush the WL and result in a reduction of WL thickness from $w + \delta$ down to δ. As mentioned in the introduction, the crushed WL material just after failure is in a fluid state F. Once initiated, the failed zone may extend under a driving "force" equal to the work of the slab weight per square meter $\rho g h$ along its vertical displacement $w_{//}$. From above definitions, this work can be written:

$$\rho g h w_{//} = \rho g \left(h_{//} \cos \alpha \right) \left(w / \cos \alpha \right) = \rho g h_{//} w \qquad (B12)$$

Under the above assumption that both $h_{//}$ and w are independent of the slope angle, so should be the driving force for WL failure.

The critical radius of the collapsing zone (in Griffith's sense (Griffith 1920)), above which it may extend spontaneously, can be easily computed as a balance between the driving force on the one hand,

and the resistant force opposing crack extension, determined by the energy required for crushing the WL, on the other hand. Such a driving force for crushing the WL was calculated under the assumption of a constant h (and not $h_{\|}$) (Heierli et al. 2008a), which might lead to fairly different conclusions than ours. However, both their driving force and ours coincide on flat terrain, where $h = h_{\|}$. Our remark above, that the driving force in the case of vertical snow fall is independent of slope angle (eq. (B12)), allows us to generalize to all kinds of slopes their results on flat terrain.

One of their results was that the critical size is fairly low, with a typical value on flat terrain of a few decimeters. This result is valid in our case for all slope angles. This means that a small local collapse (larger than a few decimeters) is likely to readily propagate along the whole WL, whatever the slope.

ii) Subsequent WL shearing and down-slope slide of the slab

The question is now whether this crushed zone would result in avalanche release or not. Since the WL is already collapsed, this second stage is driven by the shear component of the load only. Two different situations may be contemplated, depending on whether the shear stress τ in the WL is smaller or larger than the critical shear stress value $\tau^\star = \eta_o \sqrt{B \Sigma^\star / 4A}$, defined using eq. (B11), and corresponding to the value $N_c = N_c^\star$.

For sufficiently large slope angles and slab weights, the shear stress τ in the WL is larger than the critical shear stress $\tau^\star$. In this case, the strain rate is sufficiently large to maintain the WL in the fluid state, and the slab can go on sliding down.

For small slope angles and/or slab weights, the conditions are such that the shear stress is less than $\tau^\star$, and the WL clots into the S phase after a time Δt. However, during the clotting process, the collapsed zone continues to expand, leaving a fluid zone in its wake. Assuming that the collapsed zone has a circular shape, the collapsed area consists of a central disk of already clotted snow surrounded by a ribbon of still fluid granular material having almost no shear resistance. Let V be the propagation velocity of the border of the collapsed zone, which has been schematized by the propagation of a solitary wave and theoretically computed by Heierli (2005). As clotting starts at a time Δt after the passage of the wave, the width of the ribbon zone is $V\Delta t$. As a consequence, in spite of the clotting mechanism, one may wonder whether such a ribbon of fluid WL, mechanically equivalent to a shear crack, may become unstable or remain stable. This is a "Griffith-like" problem, but its specific geometry requires a particular treatment. We show in appendix C that, whatever the ribbon width, the system is always stable. The clotted zone is thus expected to eventually expand on the whole slope, probably associated with an audible "whumpf," but without any avalanche release.

We shall illustrate now these two cases in a numerical example close to typical field conditions, comparing applied and critical stresses for different slope angles. We found above (Fig. B.1) that the fluid/solid transition takes place for a critical Σ value $\Sigma^\star \approx 7.04$. However, as we do not have numerical figures for the proportionality coefficient before the exponential in eq. (B5) and for the pure "liquid" viscosity η_o involved in eq. (B11), nor any values from specifically dedicated experiments, we shall make here very crude estimates from literature data, in order to obtain no more than an order of magnitude for the critical shear stress $\tau^\star$.

The critical shear stress $\tau^\star$ has been indirectly estimated from crack-face friction experiments (Rutschblock or PST) (Van Herwijnen and Heierli 2009). In some of their experiments, the slab comes quickly to a rest (e.g. in their Fig. 4) instead of accelerating down-slope (in their Fig. 2), which is a clear illustration of a very sharp F/S transition. Indeed, their experiments C5 and C1 for instance exhibit respectively an accelerating and a clotting behavior, both for a slope angle of 33°, suggesting that both observations are made very close to the transition. With $\alpha \approx 33°$ and a slab depth around $h \approx 30$ cm (see their Fig. 1), and taking a slab density of 300 kg/m³, the transition should occur for a critical shear stress $\tau^\star \approx 480$ Pa.

Such a critical stress has to be compared with the shear stress experienced by the WL:

$$\tau = \rho g h \sin\alpha = \rho g h_{\|} \cos\alpha \sin\alpha = \frac{\rho g h_{\|}}{2} \sin 2\alpha \qquad (B13)$$

This is illustrated in Fig. B.3 for different slab depths ranging from 10 to 40 cm. The $sin2\alpha$ function responsible for the inverted U-shaped shear stress curves results from our assumption that snow falls

Appendix B: Modeling Fluid to Solid Phase Transition   75

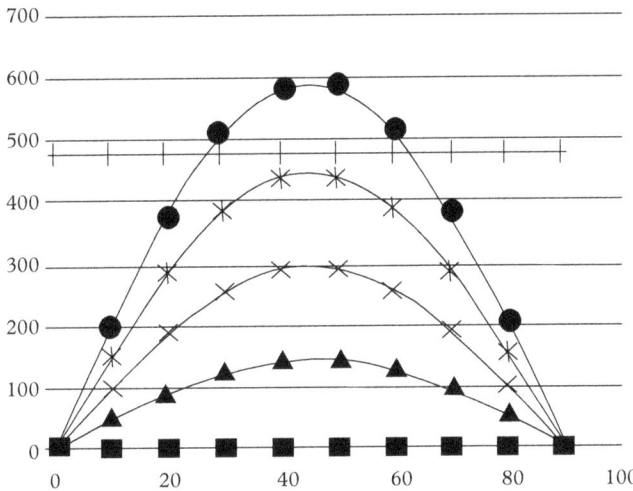

Fig. B.3 *Comparison of the critical shear stress τ^* (in Pa) with the shear stress τ experienced by the weak layer, for slope angles from 0 to 90°, and with a slab density of 300 kg/m³. Slab depths (measured vertically) are 10 cm (triangles), 20 cm (crosses), 30 cm (stars), and 40 cm (disks). In this example, avalanche release cannot occur for slab depths equal to or less than 30 cm. For a depth of 40 cm, avalanching becomes possible above a slope angle of 28°, and up to a slope angle of 62° (see text).*

vertically on average, giving a constant vertical $h_{//}$ depth. It goes to zero for slope angles of both zero and 90°, since on moderate slopes the actual slab weight is large but its shear component tends to zero, whereas on the steepest slopes the shear component is large, but the amount of snow tends to zero. The horizontal line shows the critical stress estimate $\tau_c = 480$ Pa for this particular example. It intersects the shear stress experienced by the WL for slab depths larger than about 33 cm. For a slab depth of 40 cm, avalanches can be triggered for slope angles larger than 28°, and up to 62°.

From Fig. B.3, and within the limits of our assumptions, inferences are as follows:

i) <u>Low slopes</u>: large normal-slope stresses result in WL crushing (mainly by collapse) and associated whumpfs; slope-parallel motion quickly comes to a rest due to clotting of the fluid phase, and avalanches are not released.
ii) <u>Intermediate slopes</u>: shear stresses and associated shear strain rates are sufficiently large to keep the WL in the fluid state, and the avalanche is likely to be released.
iii) <u>Steep slopes</u>: same conclusions as in i). Such a prediction may seem to be surprising at first glance, but, due to shallow slabs, shear stresses may be reduced below the critical value τ^*. Crown cracks are indeed sometimes observed to open at the junction between intermediate and very steep slopes.
iv) The question arises whether, after clotting (i.e. whumpfing without avalanche release, as in i)), the clotted WL may be destabilized again by an additional mechanical shock, as mentioned in the introduction, for instance during a skier's sharp turn or fall. However, beyond the fact that the sintered WL is significantly tougher than the initial non-collapsed one, as clotting took place for stresses lower than τ^* (top curve in Fig. B.1), a possible shear crack due to a local solid to fluid transition in the WL and experiencing the same stress would not propagate. The slope would remain safe, except possibly after another snow fall that would bring the $\psi(Nc)$ curve of Fig. B.1 from the top curve to the bottom one.

It can be concluded that, in agreement with (Van Herwijnen and Heierli 2009), our model predicts a sharp transition between whumpfing and avalanching, but in contrast with this paper, this sharp transition is not discussed here in terms of an evolution of a Coulomb crack-face friction stress between two solids, that should depend on (and increase with) normal stress, but in terms of a sudden change in viscosity during a phase transition from a fluid to a solid phase (or conversely) in the collapsed WL, controlled by the shear strain rate. By contrast with Coulomb's case, a load increase (i.e. an increasing shear stress) results here in a decrease of viscosity. The huge viscosity discontinuity between S and F phases evidenced in field experiments (Duclos et al. 2009), typically from 10 to about 10^5 Pa.s, confirms indeed that a friction analysis based on Coulomb friction is invalid in the present case.

The above results stand for the ideal case of infinite, smooth, and uniform slopes. By contrast, real terrain may show slope changes and specific boundary conditions that may affect the results. For instance, a human-triggered collapse may occur on flat terrain (e.g. a thalweg), and propagate to adjacent slopes. If snow depth and slope on the adjacent slope result in a shear stress larger than τ^* (bottom curve of Fig. B.1), an avalanche would be released from the slope, and possibly bury the people who triggered the collapse in the thalweg.

On the opposite, the shear stress computed for a uniform slope may be reduced by boundary conditions (stauchwall, gully banks, etc.). Such a decrease (and distortion) of the applied shear stress curve of Fig. B.3 shows that the critical angle for avalanche release is expected to increase, and the slope may remain stable when the distorted shear stress curve lies entirely below the critical shear stress τ^* (horizontal line).

In addition, snow transportation by wind may locally increase snow depth and also result in a distortion of the inverted U-shaped curves, in particular for large slope angles, favoring avalanche triggering on such slopes.

Finally, slab release not only requires down-slope slab shift, but also slab rupture, i.e. crown crack opening (Louchet 2000, 2001, Faillettaz et al. 2004). This last point is scarcely mentioned in slab release models. Crown crack opening takes place when the increasing tensile load experienced by the upper zone of the slab (due to the slope parallel component of the destabilized slab weight) exceeds slab tensile strength, yielding avalanche release. Despite the fact that the fluid ribbon surrounding the clotted disk is stable vs a shear crack expansion, as shown in appendix C, if its width is large enough, a crown crack may open at its junction with the clotted disk, releasing an avalanche.

In summary, the present model is based on previous field observations (Duclos et al. 2009) showing that crushed weak layers are made of a granular material that can experience sudden phase transitions from a granular fluid phase to a granular solid one, and conversely.

On this basis, evolution equations can be derived, considering shear-rate dependent erosion and aggregation kinetics of ice grains, that qualitatively reproduce the observed behavior. The model is then applied to the slab avalanche triggering problem. It predicts a sharp transition between whumpfing and avalanching, controlled by the shear strain rate of the collapsed weak layer. Within our simplifying assumptions, avalanching becomes theoretically possible when the slope angle exceeds a value typically between 10° and 40°, depending on slab depth, slab density, and slope (Fig. B.3). However, such results, obtained for infinite, smooth, and uniform slopes, are likely to be modified on real slopes, where stauchwalls or other boundary constraints may shift the critical slope angle up to higher values, in closer agreement with statistical data, as reported for instance on the "data-avalanche" website http://www.data-avalanche.org/. By contrast, snow transportation that may increase slab depth in places is likely to favor avalanche triggering. A local slope increase may also favor a tensile crown crack opening (e.g. at the junction between clotted and still unclotted zones) and destabilize the system.

Owing to its general character, this model may be used in other cases where granular slurries may undergo abrupt viscosity changes upon continuous shear strain rate changes or sudden mechanical loading. In particular, this model may account for the sudden arrest of dense avalanche flows as vanishing slopes reduce the shear strain rate beyond the critical value, or for concrete solidification in too slow spinning mixers. It may also be relevant for wet sand fluidization upon dynamical loadings, or more hypothetically for investigating permafrost stability against building construction (Crory 1982, Vakili 1991).

Appendix C
Stability of a Sintered Weak Layer Disk Surrounded by a Ring-Shaped Fluid Weak Layer Zone

> *A mathematician is a machine for turning coffee into theorems*
> Alfréd Rényi

In the slab avalanche release analysis given in appendix B, we stated that the incipient avalanche might be re-stabilized through sintering of its central zone if the shear strain rate was low enough. The present appendix aims at supporting such a statement.

The still un-collapsed ribbon surrounding the clotted central disk of the WL is considered as a shear crack, as it opposes a very small resistance for slide as compared with that of the central clotted disk. The crack stability in such a configuration is a Griffith-like problem, but its specific geometry requires a particular calculation of the critical radius.

We consider a collapsed zone of radius $R = Vt$, where V is the velocity of the solitary wave bounding the collapsed zone, and t the time elapsed since collapse initiation at point O (Fig. C.1). As mentioned in the main text, after a time Δt, the central area starts sintering. If the clotted zone has a radius r, the still fluid ring between the wave front and the clotted zone has a width $R-r$.

As in the classical treatment of the Griffith problem, we compare the variations of the stored elastic energy and the energy required for expansion of the unclotted zone equivalent to a crack (crack opening energy), experiencing a shear stress τ.

The problem being three-dimensional, we consider the torus centered in O, with an external radius R and an inner radius r, embedding the unclotted area. The radius of the circle centered on O and running in the middle of the ring is $(R + r)/2$, and the radius of the torus section is $(R - r)/2$.

Therefore, the volume of the torus centered on O, that embeds the unclotted ring, is:

$$\Omega = 2\pi^2 \left(\frac{R-r}{2}\right)^2 \left(\frac{R+r}{2}\right) = (\pi^2/4)(R-r)^2(R+r) \tag{C1}$$

and its variation for a fluctuation dr of the clotted zone radius r is given by:

$$\frac{d\Omega}{dr} = (\pi^2/4)(3r^2 - 2Rr - R^2) \tag{C2}$$

At a time t, i.e. for a given R value, the energy balance between the variations of the stored elastic energy W_1 and the crack opening energy W_2 for a fluctuation dr of the crack size around the clotted zone is:

$$\frac{dW}{dr} = \frac{d(W_1 + W_2)}{dr} = \frac{\tau^2}{2E}\frac{d\Omega}{dr} - 2\pi\gamma\, r = \frac{\tau^2}{2E}(\pi^2/4)(3r^2 - 2Rr - R^2) - 2\pi\gamma\, r \tag{C3}$$

where τ is the shear stress, $\tau^2/2E$ the corresponding stored elastic energy density, and γ the free surface energy.

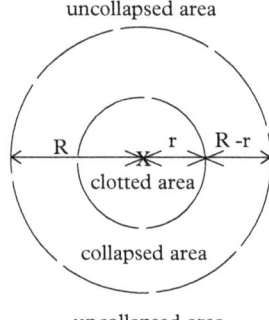

Fig. C.1 *A collapsed zone of radius R expands at a velocity V in all directions. In the center of this zone, the weak layer in a disk of radius r has sintered after a waiting time Δt.*

From eq. (C3), the critical "Griffith" radius r^* is the solution of the equation:

$$\frac{\pi^2 \tau^2}{8E}(3r^2 - 2Rr - R^2) - 2\pi\gamma\, r = 0 \tag{C4}$$

This second-degree equation obviously has two roots, one positive and one negative. The positive root r^* is given by:

$$r^* = \frac{1}{3}\left(R + \frac{8E\gamma}{\pi\tau^2} + \sqrt{4R^2 + 2R\frac{8E\gamma}{\pi\tau^2} + \left(\frac{8E\gamma}{\pi\tau^2}\right)^2} \right) \tag{C5}$$

In addition, the coefficient of r^2 in eq. (C4) being positive, $dW_1 > dW_2$ for $r > r^*$. The ring-shaped crack is therefore unstable for $r > r^*$.

However, the critical radius r^* given by eq (C5) is obviously larger than:

$$r_0^* = \frac{1}{3}\left(R + \sqrt{4R^2}\right) = R \tag{C6}$$

As r is always smaller than R, the ring-shaped crack is never unstable.

It can be noticed that if the radius of the torus section $(R-r)/2$ is large as compared to slab depth, the elastic energy would be mainly stored in the lower part of the torus (older snow). This modification would change $8E\gamma$ into $16E\gamma$ in eq. (C5), but eq. (C6) would remain unchanged, as well as the conclusions.

References

Bak, P., C. Tang, and K. Wiesenfeld, 1987. "Self-organized criticality: An explanation of the $1/f$ noise". Phys. Rev. Lett. 59: 381. doi: https://doi.org/10.1103/PhysRevLett.59.381.

Bak P., C. Tang, and K. Wiesenfeld, 1988. "Self Organised Criticality". Phys. Rev. A (38): 364.

Ballesteros-Cánovas J.A., D. Trappmann, J. Madrigal-González, N. Eckert, and M. Stoffel 2018. "Climate warming enhances snow avalanche risk in the Western Himalayas". PNAS March 27, 2018 115 (13): 3410–15; doi: https://doi.org/10.1073/pnas.1716913115.

Bartelt P., C. Pielmeier, S. Margreth, S. Harvey, and T. Stucki, 2012. "The underestimated role of the stauchwall in full-depth avalanche release". Proc. International Snow Science Workshop, Anchorage: 127.

Bazant, Z.P., G. Zi, and D.M. McClung, 2003. "Size effect law and fracture mechanics of the triggering of dry snow slab avalanches". Journal of Geophysical Research 108 (B2): 2119. doi: 10.1029/2002JB001884.

Berwin, B., 2019. "Avalanches menace Colorado as climate change raises the risk". Inside Climate News. https://insideclimatenews.org/news/08032019/avalanche-climate-change-risk-snow-storm-forecast-colorado-switzerland.

Campbell, G.A., and G. Forgacs, 1990. "Viscosity of concentrated suspensions: An approach based on percolation theory". Physical Review A 41(8): 4570.

Chaytor, J.D., S. Uri, A.R. Solow, and B.D. Andrews, 2009. "Size distribution of submarine landslides along the US Atlantic margin". Marine Geology 264 (1–2): 16.

Consiglio, R., D.R. Baker, G. Paul, and H.E. Stanley, 2003. "Continuum percolation thresholds for mixtures of spheres of different sizes". Physica A 319: 49.

Coubat G. and A. Duclos, 2017, private communication.

Crory, F.E., 1982. "Piling in frozen ground". J. of the Technical Councils of the American Society of Civil Engineers 108 (1): 112.

Daffern, T., 1992. "Avalanche Safety for Skiers and Climbers", 2nd edition. Calgary: Rocky Mountain Books.

Deschanel S., L. Vanel, G. Vigier, N. Godin, and S. Ciliberto, 2006. "Statistical properties of microcracking in polyurethane foams under tensile tests, influence of temperature and density". International Journal of Fracture 140 (1–4): 87.

Duclos, A., S. Caffo, M. Bouissou, J. Blackford, F. Louchet, and J. Heierli, 2009. "Granular phase transition in depth hoar and facets: a new approach of snowpack failure?" International Snow Science Workshop, Davos: 61. http://www.slf.ch/.

Dussauge, C., J.-R. Grasso, and A. Helmstetter, 2003. "Statistical analysis of rock-fall volume distributions: Implications for rock-fall dynamics". J. Geophys. Res. 108 (B6): 2286.

Faillettaz, J. 2003. Le déclenchement des avalanches de plaque de neige: de l'approche mécanique à l'approche statistique. PhD thesis, Grenoble University.

Faillettaz J., M. Funk, and D. Sornette, 2012. "Instabilities in Alpine temperate glaciers: New insights arising from the numerical modelling of Allalingletscher (Valais, Switzerland)". Nat. Hazards Earth Syst. Sci. 12: 2977. doi: 10.5194/nhess-12-2977

Faillettaz J., M. Funk, and C. Vincent, 2015. "Avalanching glacier instabilities: Review on processes and early warning perspectives". Rev. Geophys. 53: 203. doi:10.1002/2014RG000466, http://onlinelibrary.wiley.com/doi/10.1002/2014RG000466/pdf.

Faillettaz, J., F. Louchet, J.-R. Grasso, D. Daudon, and R. Dendievel, 2002. "Scale invariance of snow triggering mechanisms". Proc. International Snow Science Workshop, ISSW 2002, Penticton, Canada: 528.

References

Faillettaz, J., F. Louchet, and J.-R. Grasso, 2004. "Two-threshold model for scaling laws of non-interacting snow avalanches". Phys. Rev. Lett. 93 (20): 208001-1–4. doi: 10.1103/PhysRevLett.93.208001.

Faillettaz, J., F. Louchet, and J.-R. Grasso, 2006. "Cellular automaton modeling of slab avalanche triggering mechanisms: from the universal statistical behavior to particular cases". Proc. International Snow Science Workshop, ISSW 2006, Telluride: 174.

Föhn, P.M.B., 1987. "The 'Rutschblock' as a practical tool for slope stability evaluation, avalanche formation, movement and effects". Proc. of the Davos Symposium, Institute for Snow and Avalanche Research, Weissfluhjoch/Davos, 1986.

Gauthier, D., and B. Jamieson, 2006. "Evaluation of a prototype field test for fracture and failure propagation propensity in weak snowpack layers". Cold Regions Science and Technology 51 (2–3): 87–97.

Gauthier, D., and B. Jamieson, 2008. "Fracture propagation propensity in relation to snow slab avalanche release: validating the Propagation Saw Test". Geophys. Res. Lett. 35 (L13501): 1. doi: 10.1029/2008GL034245.

Griffith, A.A., 1920. "The phenomena of rupture and flow in solids". Philosophical Transactions A 221: 163.

Gubler, H., and H.-P. Bader, 1988. "A model of initial failure in slab-avalanche release". Annals of Glaciology 13, 90, doi: 10.1017/S0260305500007692.

Gutenberg, B., and C. F. Richter, 1956. "Magnitude and energy of earthquakes". Annali di Geofisica 9: 1.

Heierli, J., 2005. "Solitary fracture waves in metastable snow stratifications". Journal of Geophysical Research 110 (F02008): 1. doi: 10.1029/2004JF000178.

Heierli, J., P. Gumbsch, and M. Zaiser, 2008a. "Anticrack nucleation as triggering mechanism for snow slab avalanches". Science 321 (5886): 240. doi: 10.1126/science.1153948.

Heierli, J., A. Van Herwijnen, P. Gumbsch, and M. Zaiser, 2008b. "Anticracks: a new theory of fracture initiation and fracture propagation in snow". International Snow Science Workshop, Whistler: 9. http://issw.net/2008.php.

Heierli J., M. Zaiser, and P. Gumbsch, 2010. "Die Uhrsache von Schneebrettlawinen, der Knall im Lawinenhang". Materialwissenschaft 41 (1): 31. doi: 10.1002/pluz.201001224.

Hughes A.J., 1954. "The Einstein relation between relative viscosity and volume concentration of suspensions of spheres". Nature 173: 1089. doi: 10.1038/1731089a0.

IPCC reports, https://www.ipcc.ch/sr15/.

Jamieson, J.B., and C. Johnston, 1993. "Rutschblock precision, technique variations and limitations". Journal of Glaciology 39: 666.

Jamieson, J.B., and C.D. Johnston, 1998. "Refinements to the stability index for skier-triggered dry-slab avalanches". Annals of Glaciology 26: 296–302. doi: https://doi.org/10.3189/1998AoG26-1-296-302.

Jamieson, J.B., and J. Schweizer, 2000. "Texture and strength changes of buried surface-hoar layers with implications for dry snow-slab avalanche release". Journal of Glaciology 46 (152): 151–60. doi: http://dx.doi.org/10.3189/172756500781833278.

Johnson, B., B. Jamieson, and C. Johnston. 2000. "Field Data and Theory for Human Triggered" Whumpfs"and Remote Avalanches". Proc. International Snow Science Workshop, Big Sky: 208. http://www.issw.net/2000.php.

Jones A., B. Jamieson, and J. Schweizer, 2006. "The effect of slab and bed surface stiffness on the skier-induced shear stress in weak snowpack layers". Proc. ISSW 2006, Telluride: 157.

Kirchner H.O.K., G. Michot, and J. Schweizer, 2002. "Fracture toughness of snow in shear and tension". Scripta Materialia 46 (6): 425–9. doi: http://dx.doi.org/10.1016/S1359-6462(02)00007-6].

Libbreght K., 1999. "Snow flake science". http://www.snowcrystals.com/science/science.html.

Louchet, F., 2000. "A simple model for dry snow slab avalanche triggering". C.R Acad. Sci. Paris, Earth and Planetary Sciences 330: 821–7.

Louchet, F., 2001. "A transition in dry-snow slab avalanche triggering modes". Annals of Glaciology 32: 285.

Louchet, F., 2015. "Modeling a fluid to solid phase transition in snow weak-layers. Application to slab avalanche release". arXiv:1504.01530.

Louchet F., 2016. "Weather instabilities as a warning sign for a nearby climatic tipping point?". http://arxiv.org/abs/1609.05098.

Louchet F., and A. Duclos, 2006. "A new insight into slab avalanche triggering: a combination of four basic phenomena in series". The Avalanche Review, February. 24 (3): 13–14.

Louchet, F., J. Faillettaz, D. Daudon, N. Bédouin, E. Collet, J. Lhuissier, and A-M. Portal, 2002. "Possible deviations from Griffith's criterion in shallow slabs, and consequences on slab avalanche release". Natural Hazards and Earth System Sciences 2 (3–4): 1.

Louchet, F., A. Duclos, and S. Caffo, 2013. Modeling a fluid/solid transition in snow weak layers. Application to snow avalanche release". International Snow Science Workshop, Grenoble: 52.

Miguel, M.C., A. Vespignani, S. Zapperi, J. Weiss, and J.-R. Grasso, 2001. "Intermittent dislocation flow in viscoplastic deformation". Nature 410 (6829): 667.

McClung, D.M., 1979. "Shear fracture precipitated by strain softening as a mechanism of dry slab avalanche release". J. Geophys. Research 84 (B7): 3519–26.

McClung, D., 1981. "Fracture mechanical models of dry slab avalanche release". J. Geophys. Research 86 (B11): 10783–90.

McClung, D.M., and J. Schweizer, 1999. "Skier triggering, snow temperatures and the stability index for dry-slab avalanche initiation". Journal of Glaciology 45 (150): 190. doi: https://doi.org/10.3189/S0022143000001696.

Neukom, R., N. Steiger, J.J. Gómez-Navarro, J. Wang, and J.P. Werner, 2019. "No evidence for globally coherent warm and cold periods over the preindustrial Common Era". Nature 571: 550. doi: https://doi.org/10.1038/s41586-019-1401-2.

Pauling, L., 1933. "The structure and entropy of ice and of other crystals with some randomness of atomic arrangements". J. Amer. Chemical Soc. 57: 2680.

Perla, R.I., 1978. "Failure of snow slopes". In "Developments in Geotechnical Engineering". Edited by Barry Voight. Volume 14, Part A, 731–52. Amsterdam: Elsevier.

Powell, M.J., 1979. "Site percolation in randomly packed spheres". Phys. Rev. B 20 (10): 4194. doi: 10.1103/PhysRevB.20.4194).

Roch, A., 1966. "Les déclenchements d'avalanches". Proc. Int. Symp. on scientific aspects of snow avalanches, Davos:195.

Schulson, E.M., and P. Duval, 2009. "Creep and fracture of ice". Cambridge: Cambridge University Press.

Shapiro, L.H., J.B. Johnson, M. Sturm, and G.L. Blaisdell, 1997. "Snow mechanics. review of the state of knowledge and applications". Cold Regions Research & Engineering Laboratory (CRREL) report, August 1997.

Sigrist, C., and J. Schweizer, 2007. "Critical energy release rates of weak snowpack layers determined in field experiments". Geophysical Research Letters 34 (L03502): 1. doi: 10.1029/2006GL028576.

Sornette D., 2000. "Critical Phenomena in Natural Sciences". New York: Springer.

Sornette, A., and D. Sornette, 1989. "Self-organized criticality and earthquakes". Europhy. Lett. 9: 197.

Steffen, W., J. Rockström, K. Richardson, T.M. Lenton, C. Folke, D. Liverman, C.P. Summerhayes, A.D. Barnosky, S.E. Cornell, M. Crucifix, J.F. Donges, I. Fetzer, S.J. Lade, M. Scheffer, R. Winkelmann, and H.J. Schellnhuber, 2018. "Trajectories of the Earth System in the Anthropocene" PNAS 115 (33) 8252–9. doi: www.pnas.org/cgi/doi/10.1073/pnas.1810141115.

Tremper, B., 2008. "Staying alive in avalanche terrain". Seattle: The Mountaineers Books.

Vakili, J., 1991. "Slope stability problems in open pit coal mines in permafrost regions". International Arctic Technology Conference, Anchorage: 631. doi: 10.2118/22141-MS, http://www.amazon.com/International-Arctic-Technology-Conference-Proceedings/dp/9991136738.

Van Herwijnen, A., and J. Heierli, 2009. "Measurements of crack-face friction in collapsed weak layers". Geophys. Res. Lett. 36 (L23502): 1. doi: 10.1029/2009GL040389.

Van Herwijnen, A., and B. Jamieson, 2005. "High speed photography of fractures in weak snowpack layers", Cold Reg. Sci. Technol. 43: 71.

Weiss, J., and J.-R. Grasso, 1997. "Acoustic emission in single crystals of ice". J. Phys. Chem. B101 (32): 6113. doi: 10.1021/jp963157f.

Weiss, J., W. Ben Rhouma, T. Richeton, S. Deschanel, F. Louchet, and L. Truskinovsky, 2015. "From mild to wild fluctuations in crystal plasticity", Phys. Rev. Lett. 114 (10): 105504.

Weiss, J., T. Richeton, F. Louchet, F. Chmelik, P. Dobron, D. Entermeyer, M. Lebyodkin, T. Lebedkina, C. Fressengeas, and R. J. McDonald, 2007. "Evidence for universal intermittent crystal plasticity from acoustic emission and high resolution extensometry experiments". Phys. Rev. B 76: 224110.

Williams, M.W., C. Escobar, and D.R. Hardy, 2001. "Near-surface faceted crystals, avalanches and climate in high-elevation, tropical mountains of Bolivia". Cold Regions Science and Technology 33 (2–3): 291.

Index

airborne powder avalanche 2, 11, 26, 27
arrest 11, 45, 47, 55, 62, 76
artificial avalanche 3, 16, 27, 29, 37, 39, 51, 62

basal (plane, crack) 1, 27, 29, 36, 38, 41–44, 47, 51, 53, 55, 72
Bernal-Fowler rules 6
bifurcation 19, 20, 22, 55, 62, 65, 71
bi-percolation 12, 13, 50, 52, 56
blunt, blunting 21, 22, 53
Boltzmann 70
bridging 25, 32, 34, 37
brittle, brittleness 1, 3, 9, 11, 16, 20, 22, 30, 34, 36, 44, 55
Brownian motion 10
bulk modulus 15

cellular automata 36, 38–42, 62, 65, 67
climate 4, 57–60, 62, 64
cohesion 2, 9–11, 32, 41, 45
collapse 11, 12, 22, 25, 30, 33, 35–39, 42–44, 51, 55, 56, 60, 62–64, 69, 72–78
complex, complexity 3, 5, 58, 63
compressibility 15
correlation length 58
Coulomb 11, 22, 30, 37, 61, 76
crack 16–22, 25, 26, 29–31, 33, 36, 38, 39, 42–44, 47, 51, 53, 55, 74, 76–78
creep 3, 55, 69
crown crack 17, 19, 25, 26, 29, 36, 39, 42–44, 55, 75–76
criticality, critical phenomena 1, 4, 5, 6, 17–20, 22, 30, 34, 36, 39, 42–44, 46, 47, 51–53, 55, 56, 58, 61–65, 67, 73–78

dendrite 8, 11
domino 39, 63, 66
ductility, ductile deformation 3, 15, 20, 22, 53, 55

dynamical (loading, behavior) 1, 10, 25, 37, 43, 55, 62, 69, 76
dynamical systems (theory of) 62, 69

elastic modulus 15, 18, 34
elasticity 14, 15

facets 9
flow stress 15, 16
fluctuation 10, 58, 59, 62, 64, 72, 77
force chains 11, 12
fracture 4, 10–12, 15–17, 19, 21, 30, 38, 51
friction, frictional 4, 6, 10, 11, 14, 22–24, 30, 37, 52, 55, 61, 62, 74, 76
full depth avalanche 1, 2, 4, 11, 13, 20, 29, 45–51, 53–56, 59, 62, 64

game of life 38
granular (materials, media) 4, 9, 11, 26, 30, 31, 36, 48, 51, 55, 56, 62, 69, 72, 74, 76
green house 64
gravitational flows 1, 41, 66
Griffith's criterion 9, 17, 19, 20, 30, 33, 36, 38, 39, 43, 44, 51, 53, 55, 58, 62, 64, 65, 73, 74, 77, 78
gypsum sand 1–3

hardness 16, 32, 34, 35
hazard, hazardous 3, 4, 16, 57, 62
heal (healing, healable) 4, 36, 42, 48, 55, 56, 69
hoar (depth, surface) 9, 30, 42, 73

ice 3–6, 8–11, 17, 22, 30, 50, 52, 57, 58, 60, 62, 64, 76
incompatibility stress 17
inertia 10, 16

landslides 1, 29, 38, 41, 66
loose snow 2, 3, 42, 45–47

mitigation 1, 4, 48, 57

Navier-Stokes equations 1, 61

Pauling 6
percolation 4, 5, 11–13, 48, 50, 52, 56, 62, 71
plasticity 15
power-law 25, 29, 36, 39, 46, 61, 62, 65–67
propagation saw test (PST) 25, 30, 31, 35, 38, 42, 43, 74
pressure 5, 6, 14, 22, 37, 61

quasistatic loading 16

rockfalls 41

sand 1–3, 6, 10, 42, 45, 46, 62, 65, 76
scale invariance 27, 62, 65, 66
self-organized criticality 4, 42, 62, 65
settlement 2
shear modulus 15
sluff 2, 45, 46, 58
softening 16, 55
spontaneous (failure, release,triggering) 3, 25, 29–31, 36, 39, 40, 48, 58, 59, 64, 65, 73
stacking sequence 9
stiffness 14, 17, 34, 35, 55
strain rate softening 55
strength 15, 16, 32, 50, 52, 76
stress concentration, concentrator 11, 16, 17, 20, 21
stress-strain curve 14, 15
supercritical fluid 5
supersaturated 6, 17

temperature gradient 7–9, 59, 73
tensor 14, 16, 17, 38

tipping 58, 59, 63, 64
topology 3, 11, 62
toughness 9, 18, 20, 22, 51, 55

uniaxial 14–16
universality class 38, 67

viscosity, viscous flow 3, 55, 61, 62, 70–72, 74, 76

water molecule 6, 7, 9, 11, 17, 50, 70
water phase diagram 5, 6
wave (acoustic, elastic, solitary) 16, 43, 74

weather 3
wind slab 8, 9
whumpf 2, 31, 36, 42, 43, 74–76
work hardening 16

yield stress 15, 16, 21, 53
Young's modulus 14

Biography

Prof. Dr. François Louchet received an engineering degree at Ecole Nationale Supérieure des Mines (Nancy, France), the "Agrégation" degree in Physics, and a PhD in Solid State Physics. He has been for 30 years Professor of Condensed Matter Physics at Grenoble University (Grenoble Institute of Technology). He was also Invited Professor at Ecole Polytechnique Fédérale de Lausanne (Switzerland), Guest Scientist at McMaster University (Ontario, Canada) and at Los Alamos National Laboratories (USA), OCMR Distinguished Lecturer in McMaster, Toronto, and Kingston Universities (Ontario, Canada), invited scientist at the Balseiro Institute, Bariloche (Argentina) and at Charles University, Prague (Czech Republic). He is a member of the European Physical Society (EPS) and of the French Physical Society (SFP), and the secretary of the Data-Avalanche Association. Now retired, he keeps on with his research activity, more particularly on theoretical aspects of instabilities in Physics and Geophysics.

https://sites.google.com/site/flouchet/
francoislouchet38@gmail.com

The manufacturer's authorised representative in the EU for product safety is
Oxford University Press España S.A. of el Parque Empresarial San Fernando de
Henares, Avenida de Castilla, 2 – 28830 Madrid (www.oup.es/en or product.
safety@oup.com). OUP España S.A. also acts as importer into Spain of products
made by the manufacturer.

www.ingramcontent.com/pod-product-compliance
Ingram Content Group UK Ltd.
Pitfield, Milton Keynes, MK11 3LW, UK
UKHW062306230426
12049UKWH00006B/126